Charles Merivale

The Continental Teutons

Published Under the Direction of the Committee of General Literature...

Charles Merivale

The Continental Teutons
Published Under the Direction of the Committee of General Literature...

ISBN/EAN: 9783337022686

Printed in Europe, USA, Canada, Australia, Japan

Cover: Foto ©berggeist007 / pixelio.de

More available books at **www.hansebooks.com**

THE CONTINENTAL TEUTONS.

THE CONTINENTAL TEUTONS

CONVERSION OF THE WEST.

THE CONTINENTAL TEUTONS.

BY THE

Very Rev. CHARLES MERIVALE, D.D., D.C.L.,

DEAN OF ELY.

WITH MAP.

PUBLISHED UNDER THE DIRECTION OF
THE COMMITTEE OF GENERAL LITERATURE AND EDUCATION,
APPOINTED BY THE SOCIETY FOR PROMOTING
CHRISTIAN KNOWLEDGE.

LONDON:
SOCIETY FOR PROMOTING CHRISTIAN KNOWLEDGE.
SOLD AT THE DEPOSITORIES,
77, GREAT QUEEN STREET, LINCOLN'S-INN FIELDS;
4, ROYAL EXCHANGE; 48, PICCADILLY;
AND BY ALL BOOKSELLERS.

New York: Pott, Young, & Co.

LONDON :
WYMAN AND SONS, PRINTERS, GREAT QUEEN STREET,
LINCOLN'S-INN FIELDS, W.C.

CONTENTS.

CHAPTER I.

THE CONFLICT BETWEEN ROME AND GERMANY, FROM CÆSAR TO CONSTANTINE. . . *Page* 1

Frontiers of the Celtic and Teutonic races in North-western Europe—Cæsar's account of the Celts and of the Germans—Tacitus's account of the Germans—Outline of the Teutonic Religion from Tacitus, and its Ancient Sources—Political divisions and institutions of the Germans—Collision between the Romans and the Germans on the Rhine and on the Danube—The principal Teutonic races, the Franci, the Allemanni, and the Goths — Admission of Teutonic peoples within the Empire.

CHAPTER II.

THE ELEMENTS OF CHRISTIAN CIVILIZATION DISCOVERED BY THE GERMANS ON THEIR ENTRANCE WITHIN THE EMPIRE. 21

The Teutons become acquainted with Christianity within the Empire—Spread of Christianity in conflict with divers forms of mythology—Organization of the Church within the Empire: favoured by the persecutions — Christianity penetrates beyond the frontiers, introduced by the Christian element in the Roman armies—Widely established among the Teutons of the West in the fifth century.

CHAPTER III.

ULPHILA AND THE CONVERSION OF THE
GOTHS . . . . *Page* 33

Ulphila, the Apostle of the Goths — His early residence in Constantinople and conversion to the Faith — He invents the Gothic letters, and translates the Scriptures — Becomes bishop of the Goths in the year 341 — Inclines to Arian opinions, and subscribes to the Creed of the first Council of Constantinople, 360 — The Western Goths accept his teaching — They invade the Empire and gain the victory of Adrianople — Theodosius restores the Nicene Creed, 381 — The Ostrogoths and Visigoths, and first appearance of the Huns on the Danube — Sack of Rome by Alaric, 410.

CHAPTER IV.

SEVERINUS THE APOSTLE OF NORICUM . . . 50

Assaults and retreat of Attila — Teutonic nations in Noricum — Missionary efforts of Severinus, and influence in their conversion — The conquests of the barbarians providentially retarded — Incompleteness of their conversion.

CHAPTER V.

CLOVIS AND THE CONVERSION OF THE FRANKS 57

The Burgundions and the Visigoths thrust westward by the Huns — Attila's invasion of Gaul — Battle of Chalons — Massacre at Cologne — The legend of the Eleven Thousand Virgins — The confederacy of the Franks : the Salii and Ripuarii — Victories of Clovis at Soissons and Tolbiac—His conversion and baptism, together with the Frankish warriors—A new era in the history of the Christian Church—Gaul becomes Catholic and Roman.

CHAPTER VI.

THE CHRISTIAN MISSIONARIES IN GERMANY . *p.* 75

Nicetius, bishop of Trèves—St. Lupus, bishop of Sens—St. Aloysius, bishop of Noyon—St. Amandus, a monk at Bourges, afterwards bishop of Utrecht, followed by Columbanus and other missionaries from Ireland—The Gospel planted in Thuringia and in Bavaria—Emmeran, Rupert, Alto, and Virgilius—Christian civilization replaces the Roman.

CHAPTER VII.

ST. BONIFACE THE APOSTLE OF WESTERN GERMANY 94

The English missionaries, Winfrid or Boniface (A.D. 718) follows Willibrord in Frisia—He preaches to the Thuringians, Hessians and Saxons—Practical character of his teaching—He overthrows the oak of Thor—He is supported by many English devotees, both male and female—Growing corruption of the Church in France and Western Germany—St. Boniface throws himself at the feet of the Pope—At the death of Charles Martel his son Carloman restores the estates of the Church—The rule of St. Benedict generally accepted in Germany—The abbeys of Fulda and Gall—Martyrdom of St. Boniface in Frisia.

CHAPTER VIII.

CHARLEMAGNE AND THE FORCIBLE CONVERSION
OF THE SAXONS . . . . *Page* 115

The Saxons in Lower Germany: their three divisions, Estphalia, Westphalia, and Angaria—Their maritime habits; their religion—Charlemagne's crusade against them; their forcible conversion—The Saxon Church established by English teachers—Alcuin of York—The two abbeys of Corbie—The Knights of the Teutonic order and conversion of the Prussians and other Teutons further north in the twelfth and thirteenth centuries.

CHAPTER IX.

MORAL INFLUENCE OF THE SECULAR EMPIRE
IN THE CONVERSION OF THE NORTHERN
NATIONS 148

The Council of Nicæa—The fall of Rome—St. Augustine's "City of God"—Influence of Hellenic intelligence—Success of the Arian Opinions—Alliance of Church and State.

CHAPTER X.

INFLUENCE OF THE ECCLESIASTICAL SYSTEM ON
THE CONVERSION OF THE NORTHERN NATIONS 165

Influence of the Bishop of Rome—Pope Leo and Attila—The Holy See—The Bishops and the Church—Law of clerical celibacy—Monasticism: Schools and libraries.

CHAPTER XI.

MORAL INFLUENCE OF GOSPEL TRUTH IN THE CONVERSION OF THE NORTHERN NATIONS . . . . *Page* 184

Points of contact between Teutonic religion and the teaching of the Christian Scriptures—Legend of Balder; contest of good and evil—Reverence for women—Ideas of a future life—Moral excellence of the Gospel.

CONVERSION OF THE WEST.

CHAPTER I.

THE CONFLICT BETWEEN ROME AND GERMANY,
FROM CÆSAR TO CONSTANTINE

THE line which separated the Celtic from the Teutonic population in the north-west of Europe was ill defined, even in the time of Cæsar. That high authority, although he specifies the Rhine as the common boundary of the two races, allows that the Treviri, the Menapii, and other tribes included within the territory of the Belgæ on its left bank, were of German or Teutonic origin. It would appear that both in Gaul and Britain the Teutonic element had been gradually pressing upon the Celtic, and for ages before the Roman invasion had crept stealthily westward. On the other hand, there is reason to believe that a Celtic immigration from Gaul had at the same time been working its way by fitful inroads into the south of Germany; that the Helvetii had pushed eastward beyond the wedge-shaped district within which they were confined by the lines of the Alps and the Jura mountains; that the Boii had established themselves in overwhelming force on the northern bank of the Upper Danube, while other Gaulish tribes had penetrated into Italy and Greece, and had even crossed over into Asia Minor. We may infer, that, along with the mingling of blood considerable fusion of manners and usages had also taken place among the divers nationalities which had come thus in conflict one with another.

The Germans or Teutons, Cæsar informs us, differed in many particulars from the Gauls or Celts. They were far less advanced in all that related to material cultivation; they did not combine to dwell together in towns, nor live under civic institutions. They took no interest in agriculture, but were wont, it would seem, for the most part, to roam from one spot to another for the sake of hunting and pasturage. Their chiefs were simply the leaders they chose for each casual enterprise, nor did they recognise any priestly caste or order with spiritual authority among them. The ministers of their rude cult, if they constituted any distinct class whatever, were possessors of no sacred science, nor in Cæsar's time did they enjoy the use of letters. They had no deeper theological system to teach than the deification of material nature. The gods whose worship they recommended to the people were no other than the most obvious powers of the external world, such as light and fire, which sufficed, however, to assure the Romans that they adored Sol, Luna, and Vulcan. In default of an established theology and ritual, they had recourse, as we learn from later writers, to the practice of divination and sorcery, and were noted for the influence which they allowed female prophetesses to exercise over their imaginations. Much of the sentiment of German chivalry and romance has been traced to the respect the Teutonic races thus early paid to women, and the share they allowed them in counsel and the conduct of public affairs. Aurinia, Velleda, and Ganna have obtained a name in history as inspired leaders of the people at various crises of their early career. Nor are there wanting some other glimpses of a vein of romantic sentiment mingling with the harsher features of the primitive German people. Their manners were simple and domestic; the ties of marriage and kinship were held in especial sanctity among them; their women shared with them the hardships of warfare, cheering them to the battle, and ministering to them on the field, and acquiring thereby a right to share their counsels; though fierce and cruel to their enemies, they were hospitable and even generous to strangers from whom they had suffered no injury.

Tacitus, from whom we obtain the fullest account of the Germans of the first or second century, is influenced, no doubt, by a disposition to contrast the degenerate vices of Roman civilization with the frank and manly virtues of the barbarians; nevertheless, whatever allowance we make for this political bias, we cannot fail to recognise in his pages a genuine representation, in its broader lines, of the primitive German character. It may be remarked that he differs from Cæsar in asserting that the Germans also held Mercurius in the highest honour among their deities, and paid worship also to Mars and Hercules.

which Cæsar had represented as distinctive of the Gauls. But we may interpret this to mean that in the later period of which Tacitus speaks the Germans had advanced beyond the extreme simplicity of their earlier theology; the special worship of Mercurius, the god of wealth and traffic, seems, in the language of a Roman observer, to indicate a certain scale of progress in social cultivation. When Tacitus affirms that the gods had "denied to the Germans gold and silver," professing in his sententious way to doubt whether they had done so "in anger or in mercy," we are not to understand that this people were wholly ignorant of the pecuniary value of the precious metals, which, indeed, he elsewhere asserts them to have used in commerce, but only that no mines of gold or silver had been discovered among them. As a further indication of the moral bias in the mind of this Roman writer, we may remark that in representing the Germans as holding that the gods cannot be confined within the walls of temples, nor realized to the senses under human figures, he is evidently indulging his own imagination as a cultivated theist, and setting up the primitive barbarians of the North as supposed examples of a purer and higher philosophy to the degenerate idolaters of the ancient civilization. But as the ancient Persians erected no temples to the superior powers, but worshipped them upon the tops of their mountains, so the Teutonic peoples invoked their deities and performed their sacred rites, which included human sacrifices, in the recesses of their forests, and even called these forests after the names of the mighty beings with which they in a manner identified them.¹

To this scanty outline of the Teutonic religious system a few particulars may be added from the incidental notices of other writers of antiquity. The original primeval god of the Germans was the Tuisto or Teut of Tacitus, the eponymus of their race, to whom the nation ascribed their name, and from whom they professed to derive their personal descent. The same name is otherwise written Thoth, and is also rendered in Latin by the more euphonious appellative of Teutates. Next to the day of the sun and the day of the moon, the Germans specified Tuesday as the day of Teut. Wednesday, the fourth day, was the day of Wodan, and this god, whose name is first mentioned in the eighth century, is the most clearly identified with Mercury. He may be considered as the same also with Odin, the form which his name assumed among the Teutons of Scandinavia. Nor is it improbable indeed that Teut and Wodan are only different names for the supreme divinity, as is also Thor, the god of thunder, from whom we derive our Thursday, who seems to be in some respects identical with Odin. But whether or

---

[1] Tacitus ('German.,' ch. ix.) says that the Suevi worshipped Isis, the Egyptian divinity, but he gives no explanation of the fact, nor of the figure of a galley (liburna), which seems to have been used to represent the goddess herself, unless, as he suggests, it is to be taken as a sign that this cult was brought from beyond sea. The Naharvali, an obscure and remote northern tribe, worshipped, according to the same authority, the twin deities Castor and Pollux, under the name of Alces, and other people also had other special divinities.

not the god of the Teutons was originally the supreme divinity, the legends of their various tribes gradually effected a severance and diversity among them, as was commonly the case with other heathen peoples. Freya or Frigga, from whom we derive our Friday, is represented as a goddess, or rather as the feminine principle itself. The Suevi more particularly worshipped the goddess Hertha, or mother earth, of whom so early a writer as Tacitus makes special mention. "In an island," he says, "in the ocean (the Baltic) is a virgin grove, within which is a consecrated car covered with a robe, which none but a single priest is permitted to touch. In this vehicle the goddess is supposed to be actually present, and thus, drawn by cows, she is led forth and her priest with great ceremony. The days of her arrival and return are devoted to joy and festivity, no war is then waged, no arms are borne, till the priests conduct her back to her sanctuary, as if satiated with her converse with mortals. Thereupon the car and the robe, and the goddess herself, if you please to believe it, are swallowed up in a mysterious lake. The slaves who minister to her are swallowed up together with her. Hence the secret awe and horror of the being whom none can behold without perishing."[1]

If the number of divinities worshipped by the Teutonic tribes was slender, and their attributes few and simple; a great part of the popular religion consisted in the practice of sortilege or magical arts. Thus, for instance, the Germans pretended to pry

[1] Tacitus, 'Germania,' ch. xl.

into the secrets of futurity by examining the position of a number of twigs cast at random upon a cloth, as well as from the observation of the cries and flight of birds. Like the ancient Persians on a certain famous occasion, they commonly determined what they should do or forbear from doing by the neighing of their horses. Or they set the captive of a hostile tribe to fight with a warrior of their own, and judged by the event of such single combat what would be the issue of war between the two peoples. War, indeed, was their ruling passion, and their chiefs and leaders were chosen with special reference to their valour and fortune in the field. It was to war that their religious ceremonies seem to have generally pointed; for war only could be the main object of existence of a people who practised neither agriculture nor commerce, and possessed no acquaintance with municipal institutions. Our great authority, Tacitus, does not allow us to discern in them the most bloody traits of character which are so common to barbarian races generally; but we must remember that Tacitus, in his deep discontent with the degraded manners of his own countrymen, was a favourable witness to those whom he, perhaps, already anticipated as the heirs of future empire.

The political divisions of the great German territory were, in Cæsar's view, extremely simple; extending from the Rhine eastward, but with no definite limits, the Suevi seem to be regarded by him as the principal inhabitants of the south and centre, and, in like manner, the Cherusci of the north. Besides these, he mentions the names of certain smaller tribes with whom he came personally in contact on the Rhine; such as the Ubii, the Usipetes, and the Tenctheri. Tacitus, writing a hundred and fifty years later, and with more extensive knowledge derived from the warfare which had raged so frequently throughout this period between the Romans and the Germans in the interval, is, as might be expected, much more precise on this point. From the pages of the later historian the name of the Suevi has almost entirely vanished. It would seem, indeed, that this appellation had been at the time rather collective than specific; nevertheless, it may be remarked that it still survives in the term Suabia, which has been for centuries confined to the districts between the Rhine, the Main, and the Danube. The inhabitants of this particular region were more properly known to Tacitus and the geographers of his age as the Chatti; north of this central people they placed the Cherusci, the Chauci, the Sicambri, and the Frisii, extending, with many tribes of less note, from the Rhine to the Elbe, and the ocean. To the east were situated the Allemanni and the Marcomanni,—the All-men and the Mark or Bordermen; names which appear to indicate combinations of several tribes rather than distinct nationalities. Beyond the Elbe, and reaching still farther eastward, lay the Hermunduri, the Quadi, and others; the banks of the Oder were occupied by the Burgundians and the Vandals, races which were destined to obtain far greater notoriety in the West in later times, and to play a prominent part in the overthrow of the great southern empire, after the peoples known to Cæsar had long past out of life and memory.

The century and a half which had elapsed between Cæsar and Tacitus had not apparently developed among the Germans any taste for the life of the city. The family and clients of each chieftain herded together in villages or hamlets, around which they occupied a certain extent of land as common property, defined in primitive fashion by woods and streams; and they often left a mark or strip of vacant land between themselves and their nearest neighbours. They continued still addicted to hunting and pasture, which they preferred generally even to the rudest and easiest style of husbandry, and no man, it seems, was allowed to hold his share of the common land for more than a single year's occupation. Few, it may be supposed, cared to devote their time and labour to a cultivation thus restricted. The production of wheat seems to have been unknown among them. Oats and rye supplied them with their necessary bread, barley with beer. Their overflowing numbers were constantly kept down to the level of their slender subsistence by war and famine; but both these forces urged them to frequent change of place, and compelled them to fling themselves, sword in hand, upon the tribes by which they were surrounded. Their great national movements did not lie always in the same direction, but those which were turned against the more advanced regions of the west or south were naturally the most important in their consequences.

The collision between the Germans and the

Romans on the Rhine and the Danube was protracted through several centuries. The Germans, in the time of Cæsar, were bent on thrusting themselves into Gaul. The Roman conqueror checked this movement, and chastised it by two incursions beyond the Rhine, and, while he made no permanent impression himself on the tribes to the eastward of that boundary, the policy which he bequeathed to his successors secured his conquests from their assault for at least three hundred years. Augustus extended the defences of Gaul by planting a line of fortresses along the western bank of the great river, and by driving roads through the forests, deep into the heart of Germany, to maintain his outposts. Drusus advanced to the Saale, Tiberius penetrated to the Elbe. But these incursions were stayed and baffled, and the Romans were content in the first century to regard the Rhine and Danube as the permanent limits of their empire in Europe. Meanwhile, the Romans themselves immigrated, with many of their Gaulish subjects in their train, into the territory of the Germans to escape from the imperial taxes and conscriptions; and in the second century Trajan deemed it prudent to protect, or perhaps rather to entrap, these pioneers of southern civilization, by erecting a rampart and a palisade from the Main to the Danube; a fortification which was maintained and strengthened in the third century by Probus. Under the pressure of this peaceful invasion, the western Germans made, doubtless, some progress in the arts of social life, and cultivated an exchange of products with their southern rivals. It was not till these tribes had themselves been overwhelmed by the pressure of still ruder and more restless barbarians from the further east, that the security of the Roman territory became seriously affected.

The Roman arms continued, indeed, to make some advance for several generations along the line of the Lower Danube. Domitian, or his lieutenants, established permanent settlements in Mœsia, the modern Bulgaria, on the southern bank. Trajan extended these conquests to the Euxine, and fortified them with a wall across the neck of the Dobrudja. This victorious emperor reduced Dacia, the modern Hungary and Transylvania, to the form of a Roman province, an acquisition which his next successor, Hadrian, relinquished, though the district continued to be occupied by the Roman settlers, who were ever apt to plant themselves somewhat in advance of the imperial recruiters and tax-gatherers. The influx of these stray Romans or Italians must have been very considerable, for in no other region that was once subjected to their arms did the Latin race maintain its language so generally or so permanently. The wide territory thus occupied beyond the Danube seems to have enjoyed for some time a larger measure of tranquillity than other borders of the empire. A little higher up the frontier stream the possessions of Rome were now repeatedly harassed by the Quadi and Marcomanni, Teutonic tribes of southern Germany. The occurrence of a destructive pestilence, one of the most extensive on record, seems to have seriously weakened the permanent defences of the empire. M. Aurelius engaged in several campaigns against these formidable enemies, and held them gallantly at bay. Had he lived, he might possibly have succeeded in checking them for another generation; but his son, Commodus, took the easier course of purchasing a precarious and temporary respite from their attacks. These attacks, however, though constantly repeated, were still resisted by the most vigorous of the emperors who followed; but the practice was now gradually introduced of admitting troops of barbarians as colonists within the frontier, and arming them as subsidiary forces for defence against their own countrymen from without. Even from the time of Cæsar, indeed, it had been the policy of Rome to subsidize battalions of Gauls and Germans as auxiliaries to the legions; and this practice had been itself derived from the earlier period of the wars of the Romans with the Italians. It must be allowed that the later emperors have been somewhat unjustly stigmatized, as though, in arming foreigners in their defence, they had innovated upon the severer policy of better times. Such was not actually the case. But it was from the disuse of arms by the genuine stock of Rome and Italy, which soon became exempted from the conscription altogether, and by the ancient nobility of the city, who were absolutely forbidden to serve in the imperial armies, that the defence of the empire came at last to rest upon the provincials and the barbarian auxiliaries who were so nearly akin to them.

The barbarians had thus been introduced within the imperial frontiers in the guise of auxiliaries and defenders, and we are soon led to mark the inroads they made into the heart of the provinces. We may trace their progress about the third century, under different names, in three principal directions :—

1. We have already noticed the Suevi and the Chatti as the most comprehensive titles of the German tribes on the Rhine in the time of Cæsar and Tacitus. Further to the north we have marked the Cherusci, Chauci, Frisii, and others of less distinction. All these, however, seem to have become lost in the second and third century in the common appellation of Franci, which represented, no doubt, a widely-extended confederacy, capable of moving together under a common impulse. The Franci are supposed to have assumed their name as an assertion of their freedom from all foreign control, rather than with any reference to their descent; though in later times antiquarian chroniclers discovered for them an heroic progenitor in the person of one Francus, a reported son of Antenor, and leader of a band of Trojans from the ruins of their world-famous city. About the middle of the third century of our era this horde of barbarians effected the passage of the Lower Rhine, and extended their conquests or ravages far into the north-eastern districts of Gaul. The emperor Valerian was engaged at the time in the defence of the eastern frontier against the Persians. His son, Gallienus, associated with him in the purple, repaired to Trèves, and placed his legions under the command of an officer named Postumius, who eventually betrayed his trust, and for a moment set up a Roman empire of his own, to which Gaul, Spain, and Britain promptly attached themselves. The

provincials wanted protection, and they were prepared to give any price for it. The pretended successes of this adventurer may be ascribed to negotiation rather than to arms. If he purchased peace of the Franks, whom he boasted to have conquered, it was at the cost of giving them wide admission within the limits of his dominions. The Franci carried their inroads through the whole of Gaul, penetrated into Spain, and finally crossed over into Mauritania. At the same time they covered the Northern Sea with their vessels, for they were a maritime as well as an inland people, and employed the means they possessed of transporting themselves from shore to shore, making their attacks on many points, and withdrawing at pleasure from them. The Franci effected, indeed, no political settlements in any of these expeditions; but we may suppose that they left a considerable number of their people behind them, who became eventually absorbed into the mass of the earlier population.

2. The name of the Suevi, first known to us from Cæsar, and little noticed by Tacitus, appears again at the opening of the third century, but it is speedily lost again and merged, perhaps, in that of the Allemanni, a title which plainly indicates such another confederacy as that of the Franks, but situated in the south rather than the north or centre of Germany. These people burst upon the empire from the Upper Danube, and menaced Italy on the side of the Rhætian Alps, or Illyria, as well as Gaul from the banks of the Upper Rhine. Caracalla readily made terms with them. Alexander Severus and Maximin combated and checked them. They re-

turned again and again to the attack, and at the moment when Valerian was occupied in the East, and Gallienus in the West, they plunged into the centre of the empire, overran the plains of Lombardy, and advanced as far as Ravenna on the direct road to Rome. On the first occasion they retired before the hasty levies which were dispatched to meet them; they were defeated at Milan on a second invasion, unless, indeed, they were then bought off by the gift of ample lands in Pannonia, and the espousal of their chief's daughter to the emperor. We thus remark the establishment of two principal German peoples within the limits of the empire, where they became subject to the influence of southern civilization and to the moral and spiritual training which was making progress, as we shall see, in the minds of the Romans and provincials.

3. Beyond the Elbe and the Danube the general name of Goths is vaguely given to the Teutonic hordes which roamed the great central plains of Poland and Muscovy. This confederacy, if so it may be entitled, comprised tribes of divers names; and in pressing westward, according to the common impulse of the German races, it pushed before it the Burgundians and the Vandals. The first of these latter nations penetrated into Gaul, and obtained at an early period a settlement in the district of modern Burgundy, which they have ever since retained. The Vandals entered Gaul also; but they were of a more restless character, and not only made themselves masters of Spain, where a large province still retains, in the name of Andalusia, the memory of

their conquest, but in the fifth century occupied the northern coast of Africa, and made on one occasion, under their leader Genseric, a successful raid upon Rome itself.

Of the great Gothic race Tacitus specially mentions two branches, known to him under the names of the Gottones and the Suiones. These tribes occupied the islands of the Baltic and the opposite coasts of Sweden and Pomerania. The Suiones made themselves famous as navigators and pirates, and constituted from the earliest times the terror of the northern coasts. The Gottones were habituated, even before the Christian era, to the pursuit of commerce; they exchanged their fur and amber for the products of the South at the hands of Greek and Asiatic traders. Their name is still preserved in Gothland, a district of the great peninsula which retains the appellation of their kinsmen the Suiones or Swedes. But the arms of the Gottones had penetrated southward, even across the Danube; they pretended to be the founders of the kingdom of the Getæ, which bordered upon Macedonia. The Goths and the Getæ were, indeed, acknowledged to be the one and the same race by the writers of the early Christian centuries. The fortunes of the one and the other, if we may in any way distinguish them, were so far different, that the Getæ, planting themselves on the borders of southern culture, allowed their ruder manners to take a polish to which their northern brethren continued wholly strangers, until the time when they too, thrust forward by the pressure of other races in their rear, were invited to occupy the frontier provinces of the Eastern empire, and to raise for a time a rampart against still more formidable invaders.

We need not attempt to follow other desultory notices of the great Gothic races, mingled as they are in confusion with the Lygii, the Carpi, the Gepidæ, the Sarmatians, and the Scythians. It may suffice for our purpose to say that about the middle of the third century, contemporaneously with the incursions of the Franci into Gaul, and of the Allemanni into Italy and Pannonia, an agglomeration of German and Scythian tribes is found established, under the common name of Goths, in the plains of southern Russia. Whatever mixture of blood there may have been among them, we may believe that they were mainly of Teutonic origin. The Goths formed a combination or confederacy of the eastern Teutons, much as the Franci of the western and the Allemanni of the southern. All were actuated alike by greed of conquest and plunder, and by habitual restlessness of disposition, but all were driven forward, like the waves of the ocean, by the pressure of other restless races behind them. Before them there was plenty, hunger and starvation lay behind them. Their attacks became stronger from year to year, while the resistance of the southern empire waxed constantly weaker, though occasionally revived by the vigour of its bravest leaders, and by the precarious support which the enlistment of the barbarians themselves afforded it. But again and again the tide, repelled in one quarter, overflowed every barrier in another. When Aurelian overthrew the Goths on the Danube, they betook themselves to the Euxine. From the mouth of the Don on the Dnieper they launched a flotilla which conveyed their hosts, in the year 255, to the coast of Pontus, where they sacked the wealthy city of Trapezus in the height of its unwarlike security. In 259 they made a second descent upon Asia Minor, advancing this time from the Lower Danube, and crossing the Bosporus in their galleys. A third expedition signalized the following year, in which they made their way into Greece, and ravaged, not the islands only, but the mainland; not the plains and valleys only, but the most famous of its cities, such as Athens and Corinth, Argos and Sparta. They burnt the temple of Diana at Ephesus, then still standing in all its vast dimensions, though shorn already of its ancient splendour, the ruins of which have been only uncovered in our own day.

CHAPTER II.

THE ELEMENTS OF CHRISTIAN CIVILIZATION DISCOVERED BY THE GERMANS ON THEIR ENTRANCE WITHIN THE EMPIRE.

THUS far we have traced the inroads of the German races into the Southern empire, and we have observed how from time to time they either established themselves by force, or became admitted by compact as settlers within the limits of Roman civilization. So it was that towards the end of the third century a very large element of Teutonic blood had been infused into the body of the Roman empire, not only in the border lands of the Rhine and Danube, but throughout many regions much nearer to the centre. At Rome itself, indeed, as early as the time of Augustus, so great was the number of Germans domiciled in the city, that on the occasion of the panic which followed upon the disaster of Varus, they were required to quit it in a body, as a precaution against violence and treason. We may believe, however, that for the most part the Teutonic element of the population mingled freely with the Latin or the provincial. Under the uniform pressure of the imperial government, the various races of its subjects showed, at least in the West, little jealousy of each other, and took no pains to maintain their separate independence. They became, in fact, ultimately bound together by the moral and spiritual influences to which they found themselves subjected in common. The Romans resented the hostility of the Druids, and did not hesitate to proscribe and extinguish them altogether. Deprived of their priestly caste, the Celtic population of Gaul seems to have easily forgotten its ritual, and to a great extent even its religious traditions. The government adopted the divinities of the people into its own hierarchy by the simple expedient of declaring that they were actually identical with the acknowledged deities of Greece and Rome, and the Celtic heathens crowded the temples of Mars and Jupiter with the same blind devotion they had before lavished on Hesus and Taranis. The proscription of the Druids was a political expedient to which the Romans did not find it necessary to resort again in their other dominions in the West. Of the Teutonic cults and priesthood they evinced no jealousy. But however ample may have been the infusion of Teutonic blood into the border provinces, there seems to be little or no trace there of any influx of the religious ideas of Germany. We are rather led to suppose that such ideas hung loosely upon the minds of the wandering offshoots from the centres of Teutonic population, and that the new comers into Gaul or Pannonia were generally content to take up with such religious observances as the natives themselves presented to them. These observances were no doubt for the most part merely local and variable. The religion of the old Celtic population presented no connected system, except so far as it had become merged in the Greek and

Roman mythology. But this ancient mythology of the South had been long undermined in every quarter. Though the front which it presented to the casual observer was still imposing, from the number and splendour of its temples, the wealth, the honours, and privileges of its priesthood, and the still unbroken tradition of its immemorial acceptance, the actual faith, which is the life of religion, had been wholly lost among the more intelligent classes, and only vegetated, so to say, among the multitude with a languid and silent existence. On the other hand, new spiritual forces had sprung, and were still springing, into existence. The philosophy of the day had attached itself to new forms of Oriental theology. The worship of Isis and of Mithras had recommended itself to many cultivated and impatient seekers after truth. The creed of Judaism, with the double fascination of its austere simplicity and its profound mystery, had exercised a marked influence upon the higher society of Rome and the provinces. Above all, the solemn views of both the present life and of that which is to come, which the Gospel of Jesus Christ impressed upon all who came within its reach, had already marked it out, to such as had the heart to comprehend it, as the destined religion of the future. Christianity had undergone several persecutions in its earlier stages, but these had been for the most part local and transient. Providence, in its watchful care over the germ of divine truth, had not yet subjected it to the ordeal of a deliberate and widely-organized proscription. Much water extinguishes fire, but a little sprinkling makes it only burn the brighter, and so perhaps we may be allowed to believe that the partial trials of the second century were appointed to strengthen the vital flame which was destined to overrun the world in due season. The third century was for the most part a period of respite. During this interval the seed which had been already quickened, broke out in blossom. Both at Rome and throughout the provinces the faith of Christ received a marvellous development, and showed itself outwardly in the organization of the Church, and the local establishment of its temples, schools, and priesthood. In the third century, though still denied the privileges of what the Romans denominated a "lawful religion," Christianity could not fail to be implicitly recognised as an important social organization, holding, as it then unquestionably did, the first place as a spiritual, and not the second place as an intellectual, power in the empire. The Christian writers of the age, such as Cyprian and Tertullian in Latin, Origen and the Alexandrian Clement in Greek, may fairly enter the lists against Apuleius and Lucian, as far superior to them in vigour and manliness of expression, if inferior in the graces of a polished classical style. It would seem that Christianity was becoming the leading power of the civilized world at the very moment when the fresh blood of the northern nations began to be poured so profusely into the provinces of the empire. The disciples of the new faith were beginning to gather up a history, partly legendary no doubt, yet mainly genuine, of their own earlier career. The names of their saints and martyrs were repeated from mouth to mouth, and their stories

fed the imagination of innumerable devotees. All
that was brave and gentle in the rising generation
was attracted towards the heroes of the faith. The
virtues of the Christians, the purity of their morals,
and especially of their family life, stood in marked
contrast with the open and revolting licentiousness
of the decrepit pagan society around them.

Of the actual number of the converts in the third
century we can hardly allow ourselves to conjecture,
so conflicting are the statements or assertions of the
authors before us. Undoubtedly they bore a much
larger proportion to the pagans in the East than in
the West, in the cities than in the rural districts. At
Rome their number must have been considerable,
and they formed a compact and well-organized body,
which began to exercise some political influence. We
may believe that, in the chief places of the provinces,
such as Lugdunum, Vienna, Arelas, and Treviri in
Gaul, at Emeritu, Toletua, Tarraco,[1] and many other
cities in Spain and elsewhere, they constituted an important element in the population. Many such seats
of bishops seem to have been already founded at this
period; as at Lauriacum, in Noricum, at Eboracum,
Londinium, and Lindum[2] in the remote province of
Britain; and from these centres bands of missionary
preachers were continually going forth to convey the
tidings of salvation in Christ to the most distant
corners of every province.

The persecutions of Decius and Diocletian towards
the end of the third century subjected the Church

[1] Lyon, Vienne, Arles, Trèves, Merida, Toledo, Tarragona.
[2] Lorch, York, London, Lincoln.

to a severer trial than any of the attacks that had preceded them. Undoubtedly, the courage and discipline of the believers had been partially relaxed under the more favourable circumstances in which they had lately found themselves, while, at the same time, the assault now made upon them was more violent and more searching than any before. The disasters which had accumulated upon the Empire had excited a deeper alarm among the pagan population, and the government, in its utter perplexity, had yielded to the vain hope of reviving its strength by enforcing unity of sentiment upon all its subjects. But the experiment, now for the first time thoroughly tried, with every advantage in its favour, proved to be an entire failure. Diocletian and his imperial colleagues were forced, notwithstanding their conjoined and organized efforts, to abandon their undertaking in despair. The constancy of the Christians had completely baffled them. Then at last followed the providential reaction. Constantine was the first among the rulers of the Empire to be convinced that the faith of the Christians was the only force which could impart real unity and cohesion to the body politic. No matter how few their numbers might comparatively be, theirs was the vital spirit which alone could animate the whole. We need not suppose that this essential truth dawned all at once upon the first Christian emperor. He, too, like his associates in power, had been bred a pagan, and though he followed the example of his father before him in regarding the Christians with indulgence, it was only by degrees, and as his imagination became exalted by his political triumphs, that his eyes were opened to the Gospel as a genuine spiritual revelation; but, in granting toleration to the disciples of the true faith, he made an admirable calculation of their intrinsic power. The edict of Milan animated them with a fresh vigour, while it struck the heart of the pagans with a palsy.

From that time every further step was easy. Christianity increased daily, while paganism decreased. The multitudes who, in every society, descry from a distance the rising of a new star, and turn so promptly to its light, soon followed in the track their ruler himself had indicated. The new comers from the North, who had fewer prejudices to combat, and were more open to the influences of the humanized society around them, embraced the invitation of the Church even more readily than the fastidious pagan of the South; while in Greece and Asia the old superstitions still contended with the new ideas for the mastery, and retreated from their strongholds slowly, sullenly, and not without loud remonstrances. There is little trace of such resistance among the populations of the northern provinces, and we shall see that in the course of two or three generations, the faith of the Gospel, though imperfectly understood, had become the accepted creed of the German races wherever they were established within the limits of the empire. The steps by which this marvellous revolution was effected will now demand our more particular examination.

The origin of Christianity in the interior of Germany itself is lost in obscurity. The first ray of light has been traced indeed to the second century of our era. Justin Martyr, in his "Dialogue with the Jew Trypho," ventures to declare that there is no race of Greeks or barbarians, not even of those who "dwell in their waggons or their tents," such as the Scythians and Sarmatians, actually in the rear of the genuine German races, "among whom prayers are not made to God the Father of the world in the name of the Lord Jesus." A little later Tertullian specifies these same Sarmatians and Scythians, together with the Dacians and the Germans, as people "among whom the name of Christ is held in honour." Similar testimony is given also in the third century by Irenæus and Arnobius. It must be allowed, however, that all these statements are rhetorical in their character, and if pressed too closely, would seem to prove too much. It may be quite true that, as some of the Christian orators affirmed, there was no nation accessible to the Greeks and Romans which had not at this period received at least a call to the Gospel; but the call had in many cases been hasty and transient, and had left as yet no permanent traces of its acceptance. If large deduction is to be made from the claims of the early Christians to a wide diffusion of the faith within the limits of the empire, still more must assuredly be assumed when they speak, in their high-flown enthusiasm, of regions with which they must have had a much more slender acquaintance. Pope Innocent I., indeed, in the fourth century, expresses himself more definitely, and is entitled to greater attention; nevertheless, his testimony is naturally liable to suspicion, where he says that there is no church in Italy, or in the Gauls, comprising in that term the territories bordering on the Rhine and Upper Danube, which had not been founded by a bishop who had owed his own institution to St. Peter or his successors. This wide-spread episcopacy, however, from whomsoever it may have been really derived, soon passed the boundary of the empire. The historian Sozomen, struck by the ready conversion of the German races, explains it by the circumstance of numerous priests and bishops having been carried from the provinces into captivity among them, and having seized the opportunity to convert them by the sanctity of their lives and the miraculous powers they exhibited. Doubtless the progress of this conversion may have been accelerated by the influence of the numerous Christian soldiers who were enlisted in the armies of M. Aurelius and the later emperors, and who penetrated under their banners into the recesses of the German forests. The stories, indeed, of the Thundering Legion under Aurelius, and of the Theban Legion under Maximian, can hardly be regarded as strictly genuine; they may serve, however, to show that if there were no special battalions of Christian disciples, yet might be found many individual believers in every rank of the pagan armies. We must remember, further, that the lot of these soldiers was too often a disastrous one. Of the numbers who entered the German territory in those unlucky campaigns, few, probably, returned; and such as were not slain outright, remained as bondsmen in the homes of their enemies. Multitudes of captives, Christian as well as pagan, were carried off from the provinces after every successful incursion. Many subjects of Rome fled from the tyranny of their own governors, and preferred to cast in their lot with the barbarians whom they saw so commonly triumphant. The influx of the northern nations into the South was reciprocated, and possibly almost counterbalanced by this opposite tide of migration. Wherever the believer settled, there he preached and evangelized the natives to the best of his power; and a blessing often no doubt attended upon his efforts, however simple and unlearned he might be. Preaching indeed was a great characteristic of the time. The pagan philosophers made use of it for the inculcation of moral lessons, hardly less assiduously than the Christian ministers and believers, for the propagation of a higher spiritual knowledge. We may be sure that among the Christian captives or colonists in the northern wilderness there were some, perhaps not a few, who thus did real, though hidden, service to the cause of their Divine Master.

We must not rely, indeed, upon the genuine faith and practical piety of the strays and waifs of civilization who were thus thrown in contact with the barbarians in their native wilds. But whatever errors or moral imperfections these wandering propagandists may have carried with them to sully the purity of the Gospel which they undertook to disseminate, we may be sure that they were closely followed, wherever they made an opening, by ministers of the faith better trained than themselves, and subjected to ecclesiastical discipline, who founded Churches and established the usual forms of the Christian hierarchy among them. Hence it may have happened, that while the capital and other centres of the Empire were invaded by the Arian heresy in the fourth century, the orthodox writers, whose testimony indeed alone remains to us, affirmed that the bishops of the Gauls and Germans continued faithful to the original doctrine of the Church, and preserved a remnant at least of true believers in the midst of their distracted flocks. Athanasius, the most eminent of the defenders of the Church, stationed himself, during the period of his exile, at Trèves; and he may have assisted in maintaining the courage of the orthodox Christians throughout the fiery trial which then fell upon them. It would appear that the great Roman city of Treviri or Trèves had become at this crisis a centre of Trinitarian orthodoxy. Jerome also, a generation later, resided there for some time, and composed there some of his learned and influential works. Ambrose was born there; and Martin of Tours, who belonged to the same school of genuine churchmen, distinguished himself among them by protesting, orthodox as he himself was, against the capital penalties inflicted by one too-zealous ruler upon the Priscillianists, an offshoot of the detested Manichæans.

In the fifth century we discover, according to these authorities, that the faith of Christ was very widely established among the Teutons of the West. We may count at this time the bishoprics of Trèves, Cologne, Tongres, Metz, and Toul among the Belgic or semi-Teutonic Gauls; those of Coire, Laybach, Pettau, and Lorch (Lauriacum) in the south of Germany; of Tiburium in Dalmatia; together with others, the names of which are less distinctly specified, in Rhætia and Noricum. Of the progress of the faith in the regions of the Rhine and Upper Danube we shall speak more particularly further on; but we will first turn to the story of the conversion of the Goths, and of the development of Christianity among the people on the Lower Danube, conspicuous at a somewhat earlier period, dating from the middle of the fourth century.

CHAPTER III.

ULPHILAS AND THE CONVERSION OF THE GOTHS.

ULPHILAS, or Ulphila, the apostle, as he is commonly styled, of the Goths, was born somewhere in the extensive region comprehended under the appellation of Dacia. His name, which probably signifies a wolf's cub, attests his Teutonic affinities, but he was descended from a stock residing in the centre of Cappadocia, whence his ancestors had been carried off by the Gothic invaders in the middle of the third century. Among the Dacians, with whom his parents were domiciled, he seems to have acquired, as his vernacular language, the rude and imperfect Latin which had established itself so permanently in some portions of their country. Ulphila was born A.D. 311, being the year in which the arch-persecutor Galerius died, and in which the Christian faith was legalized by the first edict of Milan. Under the vigorous system of government organized by Diocletian the empire was enabled, notwithstanding the civil wars which drained her of her best soldiers, to keep a firm grasp of her frontier provinces. The Lower Danube was bridged at more than one point (by pontoons, perhaps, rather than by stone edifices), and the communication between the opposite banks seems to have been well established. In the year 332, Constantine, still enjoying the fruits of his victories over every rival, and firmly settled in possession of all the resources of his military rule, received the submission of Alaric, king of the Goths, beyond the river, and an embassy, which was rather a convoy of hostages, was sent to him at his new capital on the Bosporus. Among these rude natives of the North destined to receive the culture of the great southern empire and carry it back with them to their native wilds was Ulphila, then twenty-one years of age.

The youthful stranger was apt to learn, and he took a keen interest in the new world thus opened to him. During the ten years that followed he resided apparently at Constantinople, and acquired there the Greek language. It is supposed that he then first became a convert to Christianity, though, as we have seen, some rays of gospel light might already have penetrated to his home across the Danube. Indeed, a bishop of the Goths, bearing the Greek name Theophilus, had appeared in the Council of Nicæa, A.D. 325; but whether his scattered flock had lain north or south of the frontier river must remain uncertain. It was of them, or such as they, that Auxentius,[1] the contemporary biographer of Ulphila, said that they lived a hungry life of spiritual destitution from defect of preaching. The young convert was soon ordained lector, his duty being to read the Scriptures in the congregation, and this office may naturally have turned his mind to the task of translation. But the Goths, for whom he specially laboured, who spoke their primitive Teutonic language, pos-

[1] Auxentius, native of Silistria, 330—340.—MS. recently discovered at Paris.

sessed no form of writing. The first business of the translator into the Gothic tongue was to invent letters for them. The difficulty of first fixing a spoken language in letters is not unknown to our modern missionaries. The process, difficult as it is, was at this period constantly going on throughout the borders of the civilized and the barbarian world. All the Teutonic idioms were at this time unwritten; all had to be fixed in Greek or Latin characters. The undertaking of Ulphila, for which he has been so much lauded, could hardly have been original; he must have derived the idea from observation or from the experience of others. Like our modern missionaries, again, he applied to the linguistic facts before him the philological principles which he had learnt from others. Nevertheless, he may fairly enjoy the credit which is claimed for him by the general voice of antiquity, of having invented for his people "letters of their own." Ulphilas, says Philostorgius, referring to his after-life, carefully watched over his people, in many ways; but for none of his meritorious services does he so extol him as for this beneficent invention. "Ulphila," says Socrates, "discovered the Gothic letters, and by his translation of the Divine Scriptures enabled the barbarian to learn the oracles of God." These letters[1] were modified, no doubt, from

---

[1] In the Gothic alphabet of 24 letters, which he invented, were four expressing sounds unknown to the Greeks or Romans. One of these was the W, another the Th. As the Greeks are supposed to have represented by their θ the Th of our theatre, it seems probable that the Th of Ulphilas had the sound of our Thee.

the Greek as then in use, which had already become the originals of many other alphabets. In like manner the Gaulish Druids, as we learn from Cæsar, had long before adopted Greek characters to express the sounds of their own language, and these same characters, in some shape or other, have been applied to their own service by most of the later European nations. These, too, in their turn, had been derived from the Phœnician or Semitic. Of all the ancient characters, the Hebrew alone claim to represent things or natural objects, and represent, it is said, the first step in descent from the Egyptian hieroglyphics.

Ulphila had now made himself a comprehensive linguist. He wrote in the Greek, the Latin, and the Gothic languages. He is said to have composed various theological and other treatises which have been long forgotten, but his great work, worth far more than all the rest together, the Gothic translation of the Scriptures, still in a great measure exists, though in an incomplete state, and affords some valuable aid towards verifying the original text. The Codex Argenteus, as it is called, from its silver letters inscribed on the purple parchment, still substantially represents his precious labours, for it contains almost the entire Gospels, though it seems to have lost one half of its leaves since it was first discovered in the sixteenth century. This unique volume, prepared, no doubt, for the use of some distinguished personage, and accordingly written, we may suppose, with special care and accuracy, dates from the end of the fifth century, not more than a hundred years from the death of the translator. Some of the chasms in the version of the New Testament may be supplied from MSS. of later date, but of the Old Testament hardly a fragment remains. We possess, it should be noted, in this Codex quite as early and genuine a monument of the Gothic language as our Greek and Latin Codices supply in the case of any classical author. It is seven centuries older than the Scandinavian Edda, five older than the German Nibelungen Lied, three older than the poem of our English Cædmon. We may trace in it the very foundation of our own mother tongue.

In the year 341, Ulphila, being then just thirty years old, was raised from the humble rank of lector to that of bishop, through the influence of Eusebius of Nicomedia. He was attending at the time at the Council of Antioch, the same which deposed Athanasius, and expunged the word Homo-ousion from the Creed. The Gothic neophyte had attached himself to the teaching of Eusebius, which diverged in some measure from that of Arius, but came sufficiently close to it to be always popularly associated with it. The hypothesis of the subordination of the Son to the Father, taken broadly, had been for some years the dominant creed of the imperial court and its adherents, and was accordingly the more likely to recommend itself to a new convert, a foreigner, and by birth a barbarian. It continued to maintain and extend its influences throughout the Empire for a period of forty years, which embraces the whole course of Ulphila's active life; and whatever may be the exact definition of his creed, which he could not, perhaps, have precisely expressed, even to himself, it is as a disciple of Arianism that he must be regarded. The ecclesiastical historian Theodoret in vain pretended that the apostle of the Goths was perverted to heterodoxy by the juggling of a false teacher in his old age; and Sozomen declares that he subscribed to the Arian formula through "inadvertence." It is probable, however, that his preaching was directed to the inculcation of the great spiritual truths of the Gospel, with little attention to the subtle refinements of the Councils. Of his episcopate his biographer says, that, "Thus preaching and giving thanks to God the Father, he flourished gloriously for forty years in his bishopric, and with apostolic grace declared the one only Church of Christ in the Greek, the Latin, and the Gothic languages. He used to contend that one is the flock of Christ, one Lord and God, one cultivation and one building, one virgin and one spouse, one vineyard, one house, one temple, one *conventus*, or concourse of Christians, while all other *conventicula*, or meeting-houses, are not churches of God, but synagogues of Satan." He was evidently persuaded that his creed was the accepted religion of the Empire, and did not care to realize the points on which it differed from the rival dogma which it had for the time supplanted, but which was so soon about to overthrow it in turn. We may easily conceive that in those days, as in the present, such subtle differences of theoretical belief were little regarded under the pressure of practical missionary work in the borderland of Christianity and heathenism.

We do not hear of the name of Ulphila being connected with any particular city as his see. The bishops who penetrated into the realms of barbarism were wanderers, rather than resident pastors, nor, indeed, did the restless people of Gothia beyond the Danube readily congregate in any fixed localities. Ulphila was bishop of Gothia, or of the Goths. As such he continued preaching and winning souls among his people, in their own native seats, for the space of seven years. After that peaceful and prosperous interval, a great persecution arose against the new converts. "After the glorious martyrdom of many servants and handmaidens of Christ, the most holy and blessed Ulphila was driven, together with a multitude of confessors, from *Varbaricum*, the land of the barbarians, and honourably entertained by the then reigning emperor, Constantius, of blessed memory; so that, as God, by the hand of Moses, delivered His people from the violence of Pharaoh and the Egyptians, and made them pass through the Red Sea, even so, by means of Ulphila, did God set free the confessors of His Holy Son from the Varbaric land, and caused them to cross over the Danube and serve Him upon the mountains," meaning, perhaps, the Balkans, "like His saints of old."

The account of Jornandes, in his history of the Goths, a hundred years later, gives some additional colouring to the same circumstances. The followers of Ulphila, it says, under the name of Lesser Goths (Minores), settled with him in Mœsia at the foot of the mountains, a numerous people, but simple and unwarlike, abounding in cattle, rich in pastures, drinking chiefly milk. Their spiritual leader seems to have succeeded in eradicating their primitive instinct of aggression and rapine. If it be true, as we are told, that in his translation of the Old Testament he had purposely omitted the Books of Kings, as affording too much encouragement to their warlike propensities, the pious fraud may be deserving of some indulgence.

Vague as these statements are, it would seem from them that the Bishop of Gothia migrated beyond the bounds of his original circuit, under stress of persecution, and settled with a number of followers in the district of Mœsia, within the limits of the empire. There, however, the converts were received and hospitably entertained by their Christian fellow-subjects. During the time that Ulphila continued to exercise his episcopate in this new region, his learning, as well as his piety, marked him out as a fitting associate in the numerous synods which were held by Constantius. That clever and intriguing emperor regarded his ecclesiastical councils as a species of political machinery; indeed, the chief of the empire having undertaken to enforce unity of religious opinion among his people, could not but regard disobedience to the decrees of his councils in the light of political misdemeanour. In 360, Ulphila seems to have subscribed to the Arian creed of the Council of Constantinople. In 361 Constantius died, and the Church had rest from these manœuvres of the palace during the short reign of his successor, Julian, who ceased to torment it by the convening of synods which kept it in such constant agitation, while he was himself intent on resuscitating the languid energies of effete classical paganism. We may hope that the apostle of the Goths employed the interval in confirming his own people in the practical duties of their Christian profession, and in strengthening them against the perils of persecution, which might now seem to threaten them even in their place of refuge.

These clouds, however, soon passed away. The reign of Jovian which followed, shorter even than the last, gave room for no changes in the state of the Church, but the pagans of the empire were, from this time, no longer able to molest it. The era of Valentinian and Valens, from 363 to 378, was marked by bloody wars and revolutions in the valley of the Lower Danube. Valens ruled in the East, and he was not slow in employing the legion which Julian had trained for Oriental conquests in enforcing the submission of the Goths beyond that frontier river. Twice he crossed the stream on a bridge of boats and gained some considerable successes, partly by engaging one set of barbarians to massacre another. He carried off, we may suppose, both cattle and slaves by way of tribute, and obtruded upon the natives the faith of the empire; for the Church of that day was far from renouncing the supposed obligation to "compel" the adhesion of the heathens which the State had conquered for her in the field. Thus peace was concluded between Valens and Romania on the one side, and Athanaric and Gothia, or Varbaricum, on the other. But Athanaric had a rival in a chief named Fritigern, who bade for the support of the Empire by yielding his complete adhesion to the Christian faith, while the Church persisted in branding Athanaric as a heathen and a persecutor. The Goths continued to supply names to the Christian martyrology. Thus, for instance, we read of the martyrdom of Nicetas (Sept. 15), a Goth by birth, of handsome form and generous soul, who was "by no means a Goth in life and manners, nor in faith." In his early youth he had attended on Theophilus, the Gothian bishop, at the Council of Nicæa; it was not till fifty years later that he was required by the heathen Athanaric to abjure the true faith, and on his persistent refusal was subjected to cruel violence. "But the saint, through all these sufferings, ceased not to sing hymns in praise of God, and to believe in Him with all his heart. Thus witnessing a good confession to the very end, he, with many of his countrymen, received a crown of martyrdom, and gave up his spirit into the hands of God." The records of these martyrdoms make mention of the frightful holocausts which followed when the Gothic chief caused the image of his national divinity to be borne through the country on a waggon, and the Christians who refused to worship it were passed through the fire before it.

In the year 376 the turn of the heathen persecutors themselves came. The Gothic tribes beyond the Danube, over whose territory the Huns had swept under Attila a generation earlier, were now overwhelmed by a second irruption of the same savage races. The Ostro-Gothic, or East Gothic kingdom of Lithuania, were the first to suffer extinction. Their veteran chief, Hermanaric, had died in his 110th year, just before the triumph of the invaders was assured. From thence the Huns descended southward, and drove the Visigoths, or Western Goths, before them. Athanaric, after an ineffectual resistance, found a refuge in the Carpathian mountains; but the bulk of his people sought an asylum within the borders of the Empire, where so many of their kinsmen had already become domiciled, and were contributing to its defence. There were settled that little flock of the Christian Ulphila, the Gothi minores, as they are sometimes denominated, ready to welcome and protect them. This horde of immigrants was led by Fritigern, who had long set himself in opposition to Athanaric, and announced himself the patron of the believers among them. He was now further prepared to embrace Christianity in his own person. The emperor Valens required the suppliants to give up their arms and surrender themselves to the Christian faith, according to the definitions of the Council of Ariminum. Under the guidance of Fritigern on the one hand, and of Ulphila on the other, the helpless strangers accepted both conditions. The bishop of the Goths, as we have seen, made no subtle discrimination between the formulas which the rival doctors of the faith might present to him. For himself, indeed, he was already committed by his own subscription to the Arian formula of belief, and where he led the way his simple and devoted people could make no scruple in following. Some time must have elapsed while these terms, transmitted by the emperor from Antioch, were being arranged, and the impatient fugitives were still assembled, daily increasing in numbers, on the left bank of the Danube, which they had neither the courage, nor, indeed, the means to cross until the Roman flotilla was placed at their service. The women and children, it was arranged, should be first transported, together with certain distinguished personages among them, to form a number of hostages for the good conduct of the warriors that were to follow. Every precaution of fear and jealousy was taken beforehand to divide these multitudes into small detachments, and quarter them at a distance one from another. The Roman officials began to count the new comers as they arrived in order on the southern bank; but the Goths were impatient, the Romans confused or frightened, and by the time the enumerators had counted 100,000 the whole mass of fugitives broke in upon them and overran the country on all sides in reckless disorder. They retained, for the most part, their arms for their own protection, but they sacrificed a vast number of their women, whom the Romans carried off for sale in the slave-markets of Asia. Meanwhile the imperial authorities took no heed to provide for the reception of their new and dangerous guests. Lupicinus and Maximus, the prefects charged by the emperor with this essential duty, were held responsible for the disaster which ensued from their neglect. The Goths suffered the direst privations from famine. They sold their children into slavery; they were on the point of selling themselves, but Lupicinus took occasion, on hearing of their murmurs, to declare that they were plotting sedition, and attempted to seize the person of their leader Fritigern. At last the Goths rose in despair. Valens was recalled in haste from the eastern frontier to save Constantinople. The irregular warfare which ensued continued for two years. The Goths crossed the Balkans, and inflicted a severe defeat on the Romans at Adrianople. Valens, himself, perished in the burning of a cottage in which he had taken refuge. This was one of the most disastrous reverses the Roman arms had ever suffered; it was compared, at the time, to the defeat of Cannæ, but the Goths had no Hannibal to lead them, while the Romans found in Theodosius another Scipio.

This famous general, who was placed by Gratian at the head of the Eastern Empire, was, indeed, prudent as well as valiant. He was not unwilling to make terms with the assailants, in whom he recognised a valuable auxiliary force for the defence of the Empire. He enlisted their leaders in his armies, and fulfilled, in fact, the compact which Valens had suffered to be broken. The service he thus performed to the Empire was salutary and lasting. It was many centuries before its frontier recoiled from the Danube. Nor was the service he did to the Church less important; for he restored the Creed of Nicæa as the accredited formula of the Christian faith, which it has continued to be from that time to the present. Various heresies and schisms, many revolutions and reformations have left their mark on the history of Christianity, but the dogmatic definitions of the first General Council have still retained their supremacy. **The Council of Constantinople,**

A.D. 381, reaffirmed the orthodox creed, and closed the Arian controversy. The Arians had never shown much courage in the defence of their special tenets, but their sudden and complete collapse may be ascribed in a great measure to the general sense of the peril into which the empire had fallen, and its providential restoration by the arms of its orthodox defender.

It is believed that Theodosius himself sent for Ulphila, in consideration of his influence with his countrymen, to attend this Council, and help to bring about a reconciliation of conflicting dogmas, which should embrace the Gothic nation together with all the other Christian subjects of the Empire. The veteran pastor had now completed forty years of his episcopate, and was in his seventieth year, full of honours and enjoying the highest reverence among his still half-civilized people. Whatever aid he might have given to the deliberations of the Council, which was held in the spring of the year, he did not himself live to be present at it. His death occurred at the moment when the bishops were assembling, and it is happily recorded that men of all shades of opinion joined in doing honour to him at his funeral. The orthodox party, now rising in the ascendant, gladly seized on the suggestion that his heterodox views had become materially modified; but his biographer, Auxentius, records, in the work which has been quite recently discovered, the actual confession made by him in his last moments, from which the reader may judge how far he still fell short of the doctrine of the Catholic Church. It runs thus:—
"I, Ulphila, bishop and confessor, have ever thus believed, and in this alone true faith make my

testament to the Lord. I believe that there is One God the Father, alone unbegotten and invisible; and I believe in His only-begotten Son, our Lord and our God, artificer and maker of the whole creation, having none like Himself. Therefore there is one God of all, who is also God of our God. And I believe in one Holy Spirit, an enlightening and sanctifying Power, even as Christ said to His Apostles,—'Behold I send the promise of My Father upon you; but tarry ye at Jerusalem till ye shall be endued with power from on high.' And again, 'Ye shall receive power when the Holy Spirit is come upon you.' And this Holy Spirit is neither God nor Lord, but the servant of Christ, subject and obedient in all things to the Son, even as the Son is subject and obedient in all things to the Father."[1]

Having traced thus far the first conversion of the Goths in Eastern Europe, we may next turn to examine such slight indications as still exist of the progress of the faith among them in their migration westward. The presence of the Huns to the north of the Danube seems to have split the Eastern and the Western Goths into two nations, who became known to us as the Ostrogoths and the Visigoths respectively. This latter people, having already received a tincture of Gospel truth under the Arian definitions, when led by its mighty ruler Alaric, became a great and independent power, neither cowering before barbaric invasion in its rear, nor seeking favour and protection from the Empire in its front. Under Alaric the Visigoths

[1] Auxentius, cited by the *Edinburgh Review*, No. 300.

entered Italy; under Alaric, though twice bought off with gifts and tributes, they entered and sacked Rome. The effect of the conversion became at once visible. The barbarians burnt some of the Roman churches, pillaged some houses, extorted money from many captives, and outraged and murdered others; nevertheless, the records of the great catastrophe speak in the strongest terms of the general forbearance they showed; of the protection they allowed the Christian churches—especially those of Peter and Paul, the holy martyrs—to extend to fugitives, together with their most precious effects. The sack of Rome by Alaric, A.D. 210, was, on the whole, mild and merciful. While the pagans saw in this disaster the frustration of all their religious aspirations, and anticipated with the fall of Rome, the downfall of the world, the Christians quickly recovered their spirits, for the fall of the city of this world was, in their eyes, the exaltation of the city of God. The temporal power became merged in the spiritual government of mankind. St. Augustine hailed the Goths, even though they were Arians, as heralds of Christ's kingdom upon earth; in his eyes and in those of his disciples the world seemed at once to issue forth upon a new career. The old things had passed away, behold all things were become new.

The tide of Gothic invasion swept indeed back again. The hosts of Alaric disappeared behind the Alps. The form of the old Roman organization was for a time restored in Italy; but the power of Rome was broken, never to be restored; and the Church, under the rule of her bishops and popes, entered into her inheritance with spiritual issues of the gravest importance, to which we shall refer more particularly further on in another connection. The Visigoths, half Arian, half pagan, altogether barbarian, became for a time confined to the regions on the line of the river Danube, and it was there that the great effort was made for their final conversion to the orthodox faith, by Severinus, the apostle of Noricum.

CHAPTER IV.

SEVERINUS THE APOSTLE OF NORICUM.

THE great onward movement of the Huns in the fifth century was barren of direct results. The Huns were heathens of a very low type; their chief traditional object of worship was a sword, the rudest emblem of the ferocious spirit which animated them; and even this was forgotten among them till it was revived under Attila for political purposes, by the pretended discovery of the sacred weapon. Their advance into the heart of civilized Christendom, in Italy and Gaul, had little, if any, effect in turning their hearts to any form of Christianity, though the terror of their invasion accelerated the conversion of some of the barbarian nations settled peaceably in the provinces. The Christians affirmed, indeed, that Attila had been deterred from the attack on Rome by a vision of the apostles Peter and Paul; and we may well believe that both he and his warriors were overawed by the imposing appearance and language of the pope, Leo, when he ventured into their camp. Rapidly as the Huns had come to the front, they fell back not less suddenly within the limits of their trans-Danubian empire. Their conquests in Italy gained them no permanent footing in the south, while their crushing defeat at Chalons effectually broke their power of attack further westward. They had disappeared from the cis-Danubian regions of Noricum and Pannonia, which they had for a time overspread, when, in the second half of the fifth century, soon after the death of Attila, the name of Severinus makes itself known in the history of German conversion. At this period the Rugians had established themselves on the southern bank of the river, between Lauriacum (Lorch) and Faviana (supposed to be the modern Vienna); the Allemanni occupied the point where the Inn mingles with the Danube at Passau (Batava Castra); the Heruli were encamped at Juvava (Salzburg). The mongrel race of Gauls, Germans, and Goths, who still occupied these regions along with the Roman colonists, might continue to retain some walled cities and fortresses, but only to behold the heathen barbarians roaming over their fields, and carrying off their annual harvests, together with multitudes of captives. Under the calamities to which they were thus exposed, they may have forgotten or denounced the Christian doctrines in which they had been but partially instructed, and which plainly failed to secure for them mere temporal advantages. The clergy, even if true to their profession themselves, lost all influence over their flocks, which in their perplexity went, it is said, to the churches to pray, and then offered sacrifices to idols.

It was in the midst of this panic and confusion of mind that an anchorite named Severinus, a man whose birth and parentage were never ascertained, appeared in Noricum. His speech was that of a Roman, but his manners and the style of his discourses betrayed a long sojourn in the East, among the saintly hermits of the Arabian or Egyptian deserts. He had built himself a monastery before the gates of Vienna, and from thence the report of his sanctity had spread throughout Noricum. The cities which were most closely pressed by the invaders earnestly called him to their succour. He came among them over the frozen roads or the burning sands barefooted, sleeping in his hair-skin under the open sky, fasting and praying, and calling on all men to repent; prescribing the ordinances of the Christian Church as the means of a future salvation. His manners, his gestures, his utterances were a new revelation, equally to the heathen and to the degenerate Christians of the land. Their hearts were touched, not only by the sublime doctrines he preached, but by his enforcement of plain moral duties, such as the giving of alms and the ransoming of captives. He busied himself in the more worldly business of providing means of navigation for the commerce of the country, and even devised plans for its defence. By his advice the weaker posts were abandoned, and more important places strengthened against attack; the bearers of arms formed themselves at his instance into bands, and submitted to regular discipline; citizens, deserted by their magistrates, adopted a municipal organization of their own. He urged the inhabitants of Lauriacum to maintain the defence of their city against the Rugians to the last; and when their courage was at length overcome, and they were on the point of surrendering to death and slavery, he threw himself into the hostile camp, and insisted, as a messenger from Christ his Master, that the enemy should retire and leave the defenders free to withdraw in safety. The influence of the holy Severinus was also felt in many other places; and the great Roman cities on the Danube, while they fell one by one into the hands of the barbarians, effected for the most part a peaceable transfer of the laws and institutions of the empire to the keeping of more vigorous hands. The fame of the Christian apostle, both for ability and sanctity, became firmly established, and it followed as a natural consequence that the doctrines he preached were received with increasing favour.

It may easily be supposed that a preacher of such distinction was credited with more than human powers. One day a band of recruits for the Imperial service passed before him, imploring his blessing. Among them he noticed a youth of noble mien and lofty stature. "Go on," said the holy man to him; "you are clothed in wretched skins to-day, but the time will come when you will have means to enrich many. The young conscript was no other than Odoacer, the future master of Italy and of Rome, who did not forget the presage thus accorded him, but, at a later period, granted Severinus as a reward the life of a captive for whom he entreated as a token of his goodwill. On another occasion, when the Allemanni were ravaging the territory of Passau, Gibold, their king, desired to see the holy man. Severinus thereupon went to meet him, and remonstrated so boldly with him, that the barbarian promised in dismay to restore his captives and spare their country. Never, he said, had he so trembled before the face of a mortal adversary. The barbarian king Fletheus, and his Arian queen Gisa, sought an interview with him on his deathbed, and listened, not without emotion, to his earnest exhortations to practise charity and justice, and to abstain from all manner of evil. Severinus, it should be added, was a preacher of the orthodox creed, which he had learnt to appreciate in the East. No doubt the Arian heresy, derogating as it did from the divine nature of the second Person in the Godhead, had lent itself naturally in the minds of the rude barbarians of the West to paganism, and the Goths and Germans quartered in the provinces had very generally relapsed from the faith of the Gospel. It may be feared that they had in a great measure unlearnt also the distinctive precepts of Christian morality. The austerity of the new apostle's manners made a deep impression upon their simple natures; they were touched by the kindly benevolence of the man who baptized their children, healed or comforted their sick, and sent home the captives laden with presents and blessings. It is not from his disciples themselves, however, who were too ignorant of letters to record his acts, but from the Roman hagiographers, that the history of this great preacher is, not without some caution, to be gathered. Severinus impersonates, in their fond memorials, the conversion of Noricum, and may legitimately represent to us the process by which many converts and many preachers carried the truths of Christianity, step by step, through the length and breadth of a large province of the empire.

There is much truth, as well as force, in the words with which M. Ozanam[1] closes his account of this

[1] 'Études Germaniques,' ii. 42.

apostolic career. "The anchorite who defended Noricum watched, at the same time, over the interest of Christianity generally. Had the flood of the invasions rushed forward in a single tide, it would have submerged civilization altogether. The Empire lay exposed, but the nations could only enter one by one, and the Christian priesthood flung itself into the breach, so as to restrain them until the appointed moment, until, so to say, they were called by name. Attila found St. Leo at the passage of the Mincio, as he found St. Agnanus under the walls of Orleans, and St. Lupus at the gates of Troyes. St. Germanus, of Auxerre, checked Eocharich, king of the Allemanni, in the heart of Gaul, just as Severinus had restrained their warriors on the road to Italy. Posterity is not enough aware how much it owes to these noble servants of Providence, who had the glory, far from common, not of advancing their age, but of retarding it. In times so disastrous as those, ten years of delay might be the salvation of the world. If Odoacer, when master of Rome, exercised such signal clemency, if he spared her monuments, her laws, and her schools, and contented himself with destroying the empty name of her empire, it was, perhaps, because he remembered, as we have seen, the Roman monk who had predicted his victory and blessed his youthful career." Nevertheless, we must be on our guard, as we have said, against the exaggerations of the hagiographers, and observe that the conversion of the people of Noricum, whether Arians or pagans, was but partial at the time, and remained for centuries incomplete. It is allowed that six years after the death of Severinus, his disciples, still suffering under persecution, bore their master's remains in their retreat, and sought a resting-place for them beyond the Alps. The Rugians and the Heruli remained, it is said, heretics; the Allemanni pagans. Eighty years later, the Lombards descended from these regions into Italy, ascribing their conquests to the aid of Freya and Woden. At the end of the seventh century, some of these inveterate pagans still worshipped a golden dragon; and still later, the Christian clergy, in the plenitude of their power, could only wean them from this barbarous cult by placing such an image in their churches, and representing it as the brazen serpent of Moses. While the multitude clung to these gross superstitions, the chiefs of the nation long made profession of the Arian form of Christianity, from which they were at last converted, more than a century later, by the missionary zeal of St. Gregory.

CHAPTER V.

CLOVIS AND THE CONVERSION OF THE FRANKS.

THE number of Christian believers among the provincials of Gaul, during the first three centuries of the faith, was no doubt much smaller than in the eastern portion of the Empire. The conversion of Constantine, whose name was held in high honour in the quarter of the world which his father had ruled as Cæsar, may be supposed to have given considerable impulse to the propagation of the Gospel in those parts. The Church was now constituted in many dioceses throughout Gaul. Victricius, the Bishop of Rotomagus (Rouen), evangelized the natives as far as the Scheldt, preaching Christ among the Morini, whom an Italian, such as Paulinus of Nola, might still regard as the "furthest of men." A pleasant story is told of the conversion of the Marcomanni in Suabia, beyond the Rhine, through the influence of Frigitil, their queen, who had heard from a visitor from Italy of the great deeds of St. Ambrose, and was convinced herself, and succeeded in convincing her husband, of the divine favour which alone could have sustained him against his powerful adversaries. It is reported of Arbogastes, the Frankish adventurer, who won the empire with his sword, and conferred it upon the last pagan emperor, Eugenius, that he was at last converted to the faith by the assurance, whencesoever he had obtained it, that Ambrose was a being who could bid the sun to stand, and the sun stood at his command.

The manner in which, according to the ideas of the time, the invading Germans were suppose to acquire their faith in Christ, may be gathered from the quaint account of the ecclesiastical historian Socrates, in relating the conversion of the Burgundions. "I will now relate," he says, "a thing worthy to be recorded which happened about this very time. There is a barbarous nation which have their abode beyond the river Rhine; they are called the Burgundions. These people lead a quiet life; for they are, for the most part, wood-cutters, by which business they earn wages and get a livelihood. The nation of the Hunni, by making continual inroads upon this people, depopulated their country, and frequently destroyed many of them. The Burgundions, therefore, reduced to great straits, flew for refuge to no man, but resolved to entrust themselves to some god to protect them, and having seriously considered with themselves that the God of the Romans did vigorously assist and defend those that feared him, they all, by a general consent, came over to the faith of Christ. Repairing accordingly to one of the cities of Gallia, they made request to the bishop that they might receive Christian baptism. The bishop ordered them to fast for seven days, in which interval he instructed them in the grounds of the faith, and on the eighth day baptized and so dismissed them. Being encouraged thereby, they marched out against the Hunni, and were not deceived in their expectation; for the king of the

Hunni, whose name was Optar, having burst himself in the night by over-eating, the Burgundions fell upon his people, then destitute of a commander, and, few though they were, engaged and conquered very many. For the Burgundions being in number only three thousand, destroyed about ten thousand of the Huns. And from that time the nation of the Burgundions became zealous professors of Christianity."[1]

The Burgundions and the Visigoths had been propelled westward by the great wave of the Hunnic invasion of Eastern Europe, and had settled, the former apparently as orthodox Christians, the latter as Arians and semi-pagans, in the nearest parts of Gaul, when another barbarian inundation burst upon that country, embracing Sueves, Alans, and Vandals, the hordes which had followed Radagasius in his descent upon Italy, and which, after his disastrous overthrow by Stilicho, had turned all their forces against the province beyond the Rhine. They burnt the cities and slew or enslaved the mass of the population. At Moguntiacum (Mayence) they surprised the Christians in their churches, and destroyed many thousands of them in their fury, supposing them to be pagans; Christian or pagan they were themselves equally barbarians. The Huns soon followed in their track. In the middle of the fifth century, their leader, Attila, appeared in Gaul at the head of 300,000 warriors, a mingled band of many nations besides his own. Many of the chief cities of Eastern Gaul are mentioned as destroyed by this

[1] 'Socrates Scholast.,' vii. 30.

invader. But the career of Attila was first withstood at Orleans, and his mighty host was dispersed in the great battle of Chalons (A.D. 415). The Huns crossed the Rhine in their rapid flight, but they avenged their defeat by the massacre of the captives they had forced along with them as far as Cologne. The remains of this multitude of many thousands, of every age and both sexes, are supposed to be traced in some heaps of bones which have been discovered in that locality; but the imagination of the hagiographers has invented a more picturesque account of the slaughter which undoubtedly occurred. Ursula, daughter of a British chief, says the legend, had been demanded in marriage by an idolatrous prince; she consented, as the price of her father's safety, but petitioned for three years' delay. In this interval she equipped eleven galleys, and embarked in them with eleven thousand maidens of her own land, with whom she crossed the German Ocean, entered the mouth of the Rhine, and ascended the stream with a northerly wind as far as Basle. At that point the venturous troop take to land; they cross the Alps and accomplish the pilgrimage to Rome. Returning, they descend the stream again; but on their arrival at Cologne they find the Huns in possession of the place, and their further progress is arrested. Threatened with death or dishonour, they gladly sacrifice their lives; their leader, Ursula, refusing herself to share the throne of Attila. Such was the choicest version of a poetic legend, well adapted to the imagination of the times. And is it not attested to this day, ask its defenders, by the inscription found upon the spot, XI.M.V., which so plainly answers to the story of the 11,000 virgins? It is remarked, however, that these initials may just as well represent " Eleven Martyr-Virgins." Another, but a far-fetched explanation is commonly offered; namely, that the number of 11,000, or "undecim millia," reported in the chronicles, is a mistake for Undecimilla, which may have been the actual name of a single victim.

After the expulsion of the Huns, the German occupants of Gaul enjoyed some respite from foreign invasion. The Visigoths plunged farther and farther into its western regions, and finally crossed the Pyrenees, with the Vandals in their train. The Burgundions settled themselves principally on the Rhone and Saone; yet their capital was still seated at Worms, on the Rhine, at the era of Attila, as described in the Nibelungen Lied, in which the hero figures under the name of Etzel. The orthodox Church had, as we have seen, conceived at the first high expectations of these people, for on crossing the Rhine they had solicited spiritual instruction and baptism, and had received a cordial reception among the faithful, whose disciples they readily professed themselves. The nation claimed, it is said, descent from the same blood as the Romans. When the last of their bands arrived from Germany, in the year 430, it made at once a general profession of the orthodox faith. Little reliance indeed, can be placed on the loose assertions of the ecclesiastical writers in such cases as these. They must be balanced at least against other counter-statements, for we are also told that in 452, about twenty years later, the Burgundions, at Mayence, were still pagans, or had already lapsed into heresy; and had signalized their animosity to the true faith, by the murder of the bishop Aureus. At all events, it would seem that his people fell away into Arianism, or rather, perhaps, a slightly-modified paganism, before the end of the century.

It was in the middle of the fifth century that Salvian put forth his book " On the Government of God." This writer was disposed, no doubt, to take a gloomy view of all that he saw around him, and suffered few lights to mingle with the shadows which enveloped alike both the Church and the Empire. We should not, perhaps, allow ourselves to be too much appalled by the pictures of error and misery which he portrays so exclusively. On the side of idolatry he counted the Saxons, the Franks, the Gepidæ, the Alans; while every other German people, such as the Visigoths and the Ostrogoths, the Heruli, the Rugians, the Swedes, and the Vandals, were no better than Arian heretics. It was in vain, he would say, that the Church had counted upon the conversion of the Germans. It was in vain that she had taken upon herself the pleasing duty of training the unformed minds of these rude barbarians, and providing for them, through her Councils, every definition of the faith which could be rendered intelligible to their simple understandings. The disappointment of the Church was so much the greater because she had warmly boasted of the virtues of her new disciples, and bade her earlier converts to take example from them. " You think," says Salvian, addressing some of the orthodox provincials, " that you are better than the barbarians; they are heretics, you say, and we are true believers. I reply, that in faith no doubt you excel them; but in your lives, I say it with tears, you are even worse than they. You know the law, and yet disobey it; they are heretics, and do not know it. The Goths are treacherous, but they are modest; the Alans, sensual, but trusty; the Franks, deceitful, but hospitable; the Franks are dreadfully cruel, but they deserve praise for their chastity. And we profess to be surprised that God has delivered our provinces to the barbarians, although their temperance purifies the earth, deep-stained by Roman debauchery."

Such, indeed, were the complaints of the rhetorician, and his antitheses are evidently forced and unnatural. Orosius, a more sober writer, who had imbibed the spirit of his history from the broad and simple views of Augustine, in his 'City of God,' deserves much higher consideration, when he says: "Men assert that the barbarians are enemies of the State. I reply that all the East thought the same of the great Alexander; the Romans also seemed no better than the enemies of all society to the nations afar off, whose repose they troubled. But the Greeks, you say, established empires, the Germans overthrow them. Well, the Macedonians began by subduing the nations which afterwards they civilized. The Germans are now upsetting all this world; but if, which Heaven avert, they finish by continuing to be its masters, peradventure some day posterity will salute with the title of great princes those, in whom we at this day can see nothing but enemies." Augustine had indicated, not obscurely, the possibility that Providence might work out a new and grander development of humanity, from the fall of the Romans; and his disciple more distinctly points to the realization of this aspiration by the heretics, which was beginning to glimmer before his own eyes. The great triumph of the orthodox Church over the German barbarians, was at this moment at hand, and it was to be effected in the conversion of Clovis and the Franks.

This great nation or confederacy, unknown to Cæsar, and barely mentioned by Tacitus, had attained to considerable consistency in the course of the third century. The various tribes which might be comprehended under the common appellation of Franci, embraced the Sicambri, the Chauci, the Cherusci, and divers others, and extended from the mouth of the Elbe to that of the Scheldt. Two offsets from this large family become more specifically known as the Salii and the Ripuarii, both names given them apparently by the Romans; the first, indeed, is of very uncertain origin, but the last denoting, we may suppose, the residence of the tribe on the bank of the Rhine. The abode of the Salii is more distinctly ascertained. This tribe, small in number and at first of little account, occupied the Batavian island, or the tract which lies between the mouths of the Rhine and the Meuse. The nature of their position gave them advantages in defensive war, and they had obstinately maintained themselves against the forces of the empire, whose flank they commanded when it was engaged in the defence of its frontier against the German invaders. Though sternly repressed by the emperor Julian, the Salii had taken the offensive in a later generation; they had occupied the greater part of Belgium, and overreached it as far as the Somme. They had been governed by a succession of warriors whose names have been kept in remembrance by the admiration they received in a later age. From Pharamond had sprung Chlodion, from him Merovig, from him Childeric, from him Hlodwig, whom we more commonly designate as Clovis or Louis. It was at the close of the fourth century that the last-named hero, still a youth, on the decease of his father, was raised upon a buckler by his brother warriors, and acknowledged as the king or leader of the Salians. The wars to which he led them were uniformly successful. He defeated the remnant of the Roman power, under Syagrius, near Soissons; and checked an invasion of the Allemanni at Tolbiac in the neighbourhood of Cologne. By this victory he reconquered and appropriated the north-eastern part of Gaul, and thence pursuing the vanquished barbarians into their seats in Germany, he further added a still larger territory to the possessions of his national confederacy. The kingdom, or rather empire of the Franks, under Clovis as its nominal sovereign, may have ultimately embraced nearly half of Gaul and nearly half of Germany.

The Franks, so far removed from the influences of southern civilization, had as yet renounced none of their heathen superstitions. Down to his thirtieth year Clovis continued to worship the gods of his barbarous ancestors, and, under stress of circumstances, he showed a temper not less fierce and barbarous than theirs. But his temperament was not destitute, it would seem, of an element of sensibility. He became early impressed with the humane and gentle spirit of the Christian ecclesiastics with whom he came in contact, and instead of destroying their churches and ravaging their cities, he allowed them the free exercise of their rites. It happened that the Princess Clotilda, whom he had taken to wife, a daughter of the king of the Burgundions, though descended from a race of Arians, had herself been secured to the true faith by the orthodox party among them. The influence which has been often so signally exercised by the Christian missionaries in the courts of princes, and in none more powerfully, or more persistently, than in that of France, was now brought conspicuously into play. While still himself a pagan, Clovis yielded with a husband's tenderness to the mother's instances, and allowed his child to be baptized. Soon after, at the crisis of the great fight of Tolbiac, his own conscience impelled him to vow submission to the God of the Christians. "Jesus Christ," he exclaimed, as the monkish historian relates of him, "Thou whom my Clotilda declares to be the Son of the Living God, a present help to them that are in need, the giver of victory to them that believe in Thee, I implore Thy gracious aid. I have called upon my own gods for help, but they are far away from me; therefore I believe that such gods have no power to help." Nor was he slow in fulfilling the engagement he made to be himself baptized, nor remiss in bringing with him the great mass of his devoted warriors, who even, it is said, of their own accord pressed forward to offer themselves as believers. The night before his baptism, Remigius (St. Remi) sought him when he had retired from the banquet with his queen, and after long instructions and admonitions, which he was graciously permitted to give, concluded by promising them that their Christian posterity should continue to reign gloriously, should inherit the empire of the Romans, and greatly exalt the dignity of holy Church. This story is, no doubt, a striking one when read in the light of after-ages, but it is first told us by a much later writer, and must be considered as retrospective rather than prophetical.

"On Christmas-day, in the year 496, Bishop Remigius was in attendance at the doors of the cathedral of Rheims. The pavement before him was shaded by embroidered drapery suspended from the houses around. The fonts were supplied with water, and the marble floor was sprinkled with precious oils. Wax tapers of odorous scent sparkled on all sides, and the devout imaginations of the barbarians so tricked their senses, that they believed themselves to be received into the enjoyment of the perfumes of paradise. The chief of a tribe of warriors descended into the baptismal basin; three thousand of his companions followed him. And when they rose from the waters as Christian disciples, one might have seen fourteen centuries of empire rising with them; the whole array of chivalry, the long series of the crusades, the deep philosophy of the schools, in one word, all the heroism, all the liberty, all the learning of the later ages. A great nation was commencing its career in the world—that nation was the Franks."[1] There is assuredly much truth mixed up with some natural exaggeration in this patriotic exclamation of the French historian Ozanam. Let us append to it the significant cry of the new champion of the faith, when his teacher, in telling him

[1] 'Études Germaniques,' ii. 54.

the story of the Passion, dwelt piteously on the sufferings of the Divine Saviour: "Would I had been there with my valiant warriors to avenge Him." The great Christian nation which sprang from the font at Rheims has never scrupled to use the sword in the maintenance of its baptismal engagements. But the gentle Remigius had given no encouragement to this barbaric temper. His instructions had tended only to contrition and humility: "'Take the yoke upon thee," he had exclaimed, "thou subdued savage! kiss the cross thou hast ere-while burnt; burn the idols thou hast hitherto adored!"[1]

It is unquestionably true that with the conversion of the Franks commenced a new era, not only in the history of the Christian Church, but more generally in the relations of the spiritual to the temporal power, at least throughout the western parts of Europe. The authority of the Church had become, no doubt, very great in the East, and especially at the court of Constantinople. The bishops and patriarchs of the Eastern Empire were held in high honour, took part in the counsels of the government, and were employed in person as important instruments for the governing of its subjects. Constantine had affected to bow, as a humble believer, in the circle of bishops, at the Council of Nicæa. But the temporal power never subjected itself to the spiritual in the East. The throne of the Cæsars had obtained too firm a hold on the imagination

---

[1] "Mitis depone colla Sicamber; adora quod incendisti, incende quod adorasti." The story is from Gregory of Tours, who was almost a contemporary.

of the multitude even of Christian disciples in that quarter, to admit the spiritual to any equal, much less to any superior, position in relation to it. Christianity invested the emperor himself with ecclesiastical prerogatives, which might easily connect themselves in the minds of his still semi-heathen subjects with the traditions of the supreme pontificate of the pagans. But the course of religious sentiment and opinion had run in the opposite direction in the West. The abandonment of Rome by Constantine had first deprived the ancient capital of a supreme ruler, in spiritual as well as temporal things. Old Rome retained neither an emperor nor a chief pontiff. A Western emperor returned indeed for a time to Ravenna or to Milan, but never to Rome; and the great city, after a fitful period of revolutionary changes, became subject to barbarian kings of a lower type, who took no hold on her imagination. Meanwhile she had felt keenly the want of a spiritual ruler, such as she had been accustomed to venerate even in her emperors of old. This want was in due time supplied to her by the elevation of her own bishop to the spiritual supremacy. The Romans themselves were easily induced to place themselves under the ecclesiastical direction of a chief of their own choosing; but the concurrence of the Christian communities throughout the West was not all at once obtained. From the time of Leo the Great the bishops of Rome may have become conscious of the position to which they were to be called. No doubt many wise and honourable considerations concurred to urge them onward. Among others, they could not but be sensible how powerful was the influence upon the minds of the heathen barbarians before them of the unity and completeness of the papal monarchy; and the undeviating firmness with which the truths, first, of Christianity in general, and, secondly, of the orthodox creed of Christianity, had been maintained against the pagans and also against the heretics of all shades of opinion, could not fail to impress the seekers after religion with a profound awe. When a favourable opportunity occurred for striking, the Pope of Rome could deal the blow which should direct the course of ecclesiastical history for ages. Such an opportunity was afforded by the conversion of Clovis.

We have seen how the German invaders of the Empire, in the fourth and fifth centuries, had been almost universally tainted with the Arian heresy, as the first step to their conversion from their native paganism. But the inhabitants of the provinces, among whom they thrust themselves, whatever their origin, whether Celtic, Teutonic, or Latin, had adopted at this time, along with the southern civilization, the language and the ideas of the Empire, and may be contrasted generally with them by the name of Romans. The Visigoths, who had now conquered, and dominated over Gaul, were Arians; the Romans, who still formed the bulk of the population, were orthodox. The accession of the new conqueror, Clovis, to the faith of Rome, was hailed with satisfaction by the multitude, who saw in it the just vengeance of God upon the assailants from whom they had themselves suffered so much; while the Catholic, or Roman clergy, who bore spiritual sway over them, falling themselves more and more under the influence of Rome and the Roman bishop, augured from it the approaching triumph, not of blood, but of creed; not of the Roman race, but of the Roman Church. The career of Clovis continued to be a succession of bloody wars, conducted with every excess of fraud and cruelty. They are recorded for us by special biographers, men of saintly character themselves, such as Gregory of Tours, and Hincmar after him, who, regarding their hero as a Christian missionary, care not to reflect in any way upon the flagrant crimes in which he steeped himself. All they would say was, that "much must be forgiven to one who has made himself the promoter of the faith, and the saviour of the provinces." We may at least be satisfied that none of these enormities is suppressed or extenuated—the writers seem to be conscious of no motive for shame or indignation at them. It was enough for them that, when declaring war against the Visigoths of Aquitaine, Clovis should exclaim, "It grieves me much that these Arians should possess any portion of Gaul. Let us go forth with God's help and subdue them." The war, which was provoked by no injury to Clovis or his followers, but by an act of discipline exercised by the Visigoths in their own country upon one of their own bishops, was regarded as a war of religion; that in itself would have been bad enough, but it was really a contest of the clergy against the secular power; the clergy of the Frankish kingdom encouraged it and blessed it. Remigius, a man otherwise, as far as we know, of blameless character, was its main adviser and promoter. The march of the invader was attended with miracles, according to the legends, which soon grew up around it. It terminated in the great battle of Vouglé, near Poitiers (A.D. 507), which decided the contest between the Franks and the Visigoths, between the Catholics and the Arians. After the fall of the Western Empire the sovereignty of Rome and Italy had fallen into the hands of barbarians, who had only refrained from assuming the imperial title. A mysterious awe still clung to the name of empire. While all the western provinces which had once belonged to Rome had been torn away from her, and erected into separate monarchies, not one of them had ventured to seize the name of an institution which still seemed to exist, though in name only, and to exercise an influence upon the popular sentiment which could devolve upon no other. The king of Italy at this time was the Ostrogoth Theodoric. As an Arian himself, he was at enmity with the Catholic bishop; as an intruder into the ancient seat of the Roman Cæsars, he was held in no favour at Constantinople; as rival conquerors, there reigned a mutual jealousy between him and the victorious hero of the Franks. On all these accounts the emperor Anastasius was anxious to propitiate the victor of Vouglé. He conferred on him, as though by an act of sovereign authority, which he did not actually possess, the honorary title of Consul of Rome; and conveyed to him its insignia by the hands of a special ambassador. In the basilica of Tours, before the tomb of St. Martin, in the presence of priests and warriors, the long-haired leader of the Franks was invested with the purple tunic and mantle, and a diadem, and then mounting his horse, paced solemnly through the streets, casting gold and silver among the multitude. From this time his own Franks were proud to hail him as consul and Augustus. He was invited to send the younger scions of his own family to assume offices in Rome, and assist in the defence of the city whose ancient tradition he upheld in his own person. The actual sovereign, Theodoric, seems to have acquiesced in an arrangement which, while it conferred no real power upon his rival, might, in process of time, elevate him to a high position in the affections of the people there. But the descendants of Clovis were not destined to succeed to the empire of Rome, a consummation reserved for Charlemagne three centuries later. Meanwhile, the merely titular majesty of the old Roman magistracies bore some solid fruit in the consolidation of the Frankish sovereignty over Gaul, which now extended from the Rhine to the Garonne, and included the latest cessions of the Ostrogoths on the borders of the Mediterranean. When, towards the middle of the sixth century, the emperor Justinian seized by a momentary victory the sovereignty of Italy, the Arian heresy of the Ostrogoths became rapidly extinguished in the Peninsula; the whole of Gaul was now reputed Catholic, under the increasing authority and influence of the Roman Church; and, finally, the kingdom of the Franks, acknowledged as an independent sovereignty by the emperor, became the bulwark alike of Southern faith and Southern civilization against the barbarians of Germany.

The divisions of the Merovingian princes among themselves, and the disintegration of their vast territories after the death of Clovis, we need not here consider. We have followed to the end the invasions of the barbarians, and their incursions within the Empire, which led the way to their final conversion. From this time such invasions ceased. The supremacy of the Frank and of the Church was established as far as the bank of the Rhine, and even some distance beyond it. Clovis and his successors occupied large portions of Germany, wherein they continued to wage desolating wars, which contributed at least to the propagation of the faith, such as they held it, within the heart of the German territory, and to the establishment of the Church with the organization it derived from its close connection with the Roman see. From this time we must contemplate the Church of Christ under a new and painful transformation; for she becomes, with the Merovingian court for her stanch ally, an aggressive and a crusading power. The wars of religion have already commenced, wars of persecution are about to follow.

But before we are summoned to review the painful circumstances attending this new phase in ecclesiastical history, we may turn our eyes upon the pleasant picture of the missions of peace and love which still for a season carried the Gospel by private hands into the recesses of Germany.

CHAPTER VI.

THE CHRISTIAN MISSIONARIES IN GERMANY.
ST. COLUMBAN AND THE IRISH MONKS.

THE modern history of Western Europe may be said to commence from the establishment of the Frankish kingdom. The France of Clovis and his successors took up the civilization of the ancient world at the point where it had fallen from the dying hands of imperial Rome. She took up the tradition of literature, art, and science; debased and decrepit as these had now become, they did not, for a time at least, fail to recover some strength and even newness of life. She took up the tradition of Roman jurisprudence, readjusting the laws and institutions of the barbaric races, and constituting an organized civil government, founded upon the ancient lines, but guarded by the higher sanctions of the Christian faith. Above all she took up the tradition of an ecclesiastical system incorporated in the civil constitution of the State. She welded the secular and the spiritual powers together, as closely as of old, but upon different principles. For she kept the administrators of the State apart from those of the Church, the laity from the clergy; she controlled the nominal independence of the king by the check of spiritual influences, which it was always difficult, and often impossible, for him to overcome. Practically, Church and

State worked, for the most part, harmoniously together, lending each other mutual support, while the independence and eventual supremacy of the Church were actually secured by the restriction of priestly celibacy, which hindered its ministers from degenerating into a caste, and kept its aims and objects at a higher level than those of the rival secular power. The authority of the Church was maintained by its internal organization ; by its army of priests quartered in every town and village, by its official hierarchy of bishops, who ruled its districts as their provinces, and held active personal communication one with another ; by its frequent synods for more general deliberation on its vital interests ; above all, by the ever-increasing awe with which its vigour and genuine devotion to its objects could not fail to inspire the people, whom it held in spiritual subjection.

Gregory of Tours, the historian of this period, has drawn the portrait of one of the bishops, who may be taken, perhaps, as a favourable representative of the class. St. Nicetius of Trèves attained to that see in the year 527. His special characteristic was the horror he evinced for the violence of which he must have been so frequently a witness. The day that he went to take possession of his see, his attendants, on arriving near the gates of the city at sunset, proceeded to unyoke their horses and let them loose among the fields of the peasantry. He required the animals to be removed from the poor people's pastures under threat of refusing the communion to the evildoers. When these refused to obey him, he went himself in pursuit of his cattle, and drove them before him into the city amidst general acclamations. But, with all this simplicity of character, he was not insensible to the ambition of reconstructing the ancient capital of Gaul in all the grace and splendour for which it had been famous before it had been four times stormed and plundered by various assailants. He was still mastered by the ancient traditions of the empire, and it was in this spirit that he restored not only the churches but the walls and towns of Trèves. At the same time he laboured passionately for the edification of the barbarians; not, indeed, by preaching personally to assembled multitudes, but by disposing the hearts of princes and princesses as the readiest instruments for the conversion of their people. Some he praised and doubtless flattered; but others he did not spare from any fears on his own account. From one of the Merovingian kings he suffered banishment; but he was restored by another; and he ended his days attended by the veneration of the multitude, who believed that he was informed by God himself of the designs of Providence for the future career of their royal race.[1]

Nicetius, it has been said, did not exert himself to go abroad among the pagans and labour personally for their conversion in their own homes. But he was succeeded by a series of missionaries, whose calling formed a new epoch in the evangelization of the northern nations, and requires our particular notice at this period. The zeal which now, apparently for the first time, impelled so many brave and ardent spirits to encounter these new hardships and perils

[1] Ozanam, 'Études Germaniques,' ii. 81.

may have been due to some natural reaction from the spirit of reserve and solitary meditation which had gradually covered the face of Christendom with the secluded retreats of monks and anchorites. The restraints of monastic life, which had been accepted with fervour, and borne at least with resignation by the moody contemplative Orientals, pressed, it would seem, more sorely upon the active practical genius of the West, especially, we may believe, upon those in whose veins still ran the warm blood of the German barbarians, their ancestors. However this may be, it seems that from these contemplative retreats issued, not all, but many of the daring and enduring missionaries who have given a special character of their own to the sixth and seventh centuries of Christianity.

The chronicles of the Church abound in names of the missionaries of this period who attained to eminence among their fellows, and were honoured with the designation of saints. The principal incidents of their lives, the works they performed, and the miracles by which their superior holiness was supposed to be attested, are carefully recorded. It may be well to mention a few of them, as representatives of a class which doubtless effected a great work, and must ever hold a high place in the history of Christendom. St. Lupus, whom we first introduce, was not the inmate of a monastery; he was bishop of Sens, in France, but from thence he was expelled, it seems, by a combination of his own clergy with the laity against him. Driven from this episcopal home, he wandered forth, as a solitary and exile, in the early years of the seventh century, and betook

himself to the low countries of the Scheldt and
Meuse, in which we may suppose the squalid rem-
nants of the barbarian population were still devoted
to the rudest forms of Teutonic paganism. Lupus
was shortly followed, and in nearly the same regions,
by St. Aloysius (Eligius or Eloi), and Amandus. The
first of these had been an artificer in wood or stone,
and he had acquired great consideration in the
courts and councils of the Frankish kings of Austrasia.
He had been promoted to the ministry of the Church,
and had obtained, no doubt by royal favour, the
bishopric of Noyon in 640. Dissatisfied with the
too easy life which, in the general relaxation of
morals, had become prevalent among his order, he
tore himself away from the temptations to indolence
which surrounded him at his episcopal station, and
flung himself vehemently into the rude and still
almost trackless plains of Gueldres and Friesland,
where he preached the Gospel to the untutored
heathens. These barbarians had never heard the
name of Christ, and never seen a preacher or missionary of the faith. They were disposed to resent
the intrusion of the stranger; but he succeeded in
soothing their ferocity, and gradually induced them
to hearken and to obey. Every year, it is said, at the
holy Paschal season, he was wont to baptize great
numbers of them together. The old men renounced
their errors, and listened with rapture to the promises
he held out; while the young allowed and encouraged
him to declare to them the purest lessons of Christian
grace, and require them to accept them as necessary
to salvation. It may be worth while to repeat, as a

sample of the Christian preaching of the day, the sermon which St. Ouen, one of this teacher's disciples, has put into his mouth. "Worship not," he would say, "the heavens, nor the stars, nor the earth, nor anything else but God; for He, by His power alone, has created and disposed all things. Doubtless the sky is lofty, the stars are beautiful, the earth is vast, the ocean boundless, but He who made all these is greater and fairer than they. I declare, then, that you must not follow the impious customs of the unbelieving pagans. Let no man take note of what day he leaves his house, or what day he returns there, for God has made every day. Nor must any one scruple to begin a work at the new moon; for God has made the moon, to the end that it should mark the time and enlighten the darkness, and not that it should interrupt men's business and disturb their minds. Let none believe himself subject to an appointed destiny, to a lot or to a horoscope, according to the common saying, Every man shall be that which his birth has made him; for God wills that all men should attain salvation and arrive at a knowledge of the truth. But on every Sunday present yourselves at the church, and when there take no thought of business or of quarrels, or of trifling conversation, and hearken in silence to the divine teaching. It sufficeth not, my friends, to have received the name of Christians if you do not the works of Christians. That man bears the name of Christian with profit to himself, who keeps the precepts of Christ, who steals not, who bears not false witness, who lies not, who doth not commit adultery, who hateth no man, who returns not evil for evil. That man is a Christian indeed who puts no faith in phylacteries nor other devilish superstitions, but hopes in Christ only; who receives the wayfarer with gladness, as though he were entertaining Christ Himself, for it is said, 'I was a stranger, and ye took Me in.' That man, I tell you, is a Christian who washes the feet of his guests, and loves them as dear kinsmen, who bestows alms to the poor according to his own means, who touches not the produce of his own farm till he has given a portion to the Lord, who knows not the deceitful scale or the false measure, who lives chastely and in the fear of God, who finally, bearing in mind the Creed and the Lord's Prayer, takes care to teach them to his children and his household." The simplicity of such elementary Christian teaching may naturally strike us, not so much in itself, for we well know that such elements are the true food of the ripest Christian as well as of the rudest novice, but from the circumstance that it should have been thought important to record it. But, in fact, no teaching could have been more novel or more striking, or if boldly enforced, more effective in its application to the carnal, sensual, crafty barbarians of the German forests, who acknowledged among themselves no object but selfish gratification, no principle of constraint but selfish fear. "Let him rise early who seeks the life or riches of his neighbour: the wolf who keeps his lair seldom finds a prey." Such was the teaching of the Edda; such the maxims of savage life to which the sermon of Aloysius was the reply of the Christian.

St. Amandus, a native of Aquitania, and as we may suppose from his name, of the Latin race, had been bred, from early years, under monastic discipline on one of the solitary islands of the Gallic coast; at a later period he had transferred himself to the central city of Bourges, and there, too, he had passed fifteen years in a friar's cell. Weary of this long imprisonment, and with an imagination exalted by long watchings and self-inspection, he had felt a call to rush off to Rome and visit the shrines of the apostles. The warden of the holy spot had forbidden him to practise a night-watch before the tombs of the sainted martyrs; but he had seated himself, notwithstanding, on the steps which led to the basilica, when St. Peter himself seemed to appear to him in a vision, and point to the road towards Gaul, commanding him to carry the Gospel to the pagans in that direction. This summons he promptly obeyed; this manifest call was readily accepted, and he received episcopal ordination, with no defined local jurisdiction. He commenced his labours in the country about Tournay; but so dismayed was he at the brutal rudeness of the heathen barbarians around him, so disconcerted by their obstinate resistance to his teaching, that he stooped to solicit Dagobert, the king of Austrasia, to interpose with the royal authority and compel his hearers to submit to the rite of baptism. This interference only made matters worse: from remonstrance the pagans betook themselves to violence, in which the women were most conspicuous; and the apostle was deserted even by his own few companions. He remained, however, firm at the post he had chosen for his exertions, and we may believe that he ultimately made some impression upon the people by the exercise of true Christian charity. It would seem that, being unable to save a condemned criminal from punishment, he contrived to restore his life after the apparent execution of the sentence. However the restoration may have been effected, it was, of course, noised abroad as a miracle, and the pagans were so profoundly impressed by it, that they suddenly overthrew the altars of their idols and rushed in crowds to receive Christian baptism. Such is the story we have had handed down to us, which it is necessary to relate in order to put the modern reader on his guard against the superstitious legends which too often obscure the true history of these conversions; and, we may add, which vitiate and disgrace it.

The success of Amandus seems now, from whatever cause, to have been rapid and extensive. He is said to have planted the Christian faith deep in the soil of Flanders by the foundation of numerous monasteries, which he filled with neophytes whom he had ransomed from captivity, before he attached them to his preaching. Among these are mentioned Saints Bavon, Floribert, and Humbert, all known in the early Christian annals of their country, and whose names are perpetuated in many churches existing in modern times. Amandus became himself Bishop of Trajectum, or Maestricht, but he did not long continue to reside there. He had acquired all the restlessness of the true missionary spirit, and again went forth at a later date to prosecute the work of evangelizing the heathen further afield. The same roaming course was pursued, it seems, by others of the Latin missionaries, who had taken so vigorously to the work of conversion. Judging from the names of the bishops who succeeded to their sees towards the end of the seventh century, it would seem that the places of these first founders of the German Church were very commonly filled by the native race, whom they had themselves conquered for its service. But these Germans, it is said, were not so generally men of the simple, missionary spirit. They had inherited more of the warlike and secular character of their pagan fathers, and began to introduce the state of temporal barons, with their horses and dogs, their retainers, and their armed followers. They were willing to receive their appointments from the Merovingian monarchs, who found them useful auxiliaries in the field no less than in the council. It was only on the debatable frontier land of the Christian and the Pagan that religion was preached pure and undefiled, even at this early period. The episcopate, together with all the higher orders of the Church, would have soon assumed the character of a secular, or even a military caste, with hereditary succession, but for the resolute resistance of the Pope of Rome, with all the powers which circumstances threw at that time into his hands, and for the strict requirement of celibacy which he persisted in enforcing upon the clergy.

The names which have hitherto been brought before us are those of the Latin race in Gaul, who undertook the evangelization of the pagan kinsmen of their own associates in their common country. But the ranks of these valiant missionaries were strongly reinforced from another and a distant quarter. The establishment of the faith among the Celtic population of Ireland in the fourth and fifth centuries does not come within the scope of this volume, nor shall we delay to remark upon the energy with which its ministers threw themselves in the first instance upon the opposite coast of Scotland, and effected to a great extent the conversion of the Caledonians. The apostle of Scotland was St. Columba, whose name is still perpetuated in the I-Colmkill, the island of Columba of the Churches, among the Hebrides. About fifty years later, early in the seventh century, the same name, with a slight variation,—which has caused some confusion in the accounts of modern writers,—viz., Columbanus or Columban, introduces us to another Irish missionary, who first frequented the court of the Frankish kings, and when he had excited the animosity of the wicked queen Brunehault, the Jezebel of ancient France, betook himself to the German pagans on the outskirts of the Austrasian dominions. The direction which Columban took was towards the south. He ascended the Rhine as far as the Lake of Constance, and founded a monastic establishment at Brigantium, or Bregenz. Every step he took is profusely marked by legendary accounts of miracles or visions, which only serve to confuse and disfigure the references they contain to the actual existing superstitions of the heathens among whom he thrust himself. It would seem, however, that the Teutonic people of these obscure regions were still addicted to the simple rites of their ancestors, who so many centuries before had worshipped the three great German divinities, Donar, Woden, and Saxnot, equivalent to the Scandinavian Thor, Odin, and Freya.

This worship had invaded a church which, at one time had been dedicated to a Christian saint, Aurelia, but which had been now profaned by the barbarians, and adorned by them with golden images affixed to its walls. Paganism, it would seem, here as elsewhere, was constantly encroaching upon the Christianity which had gained over it from time to time a transient or an apparent triumph. We must, indeed, bear in mind that throughout the early ages of conversion the advance of the true faith was subject to repeated reverses, nor must we be surprised to find the missionaries of one age engaged in the evangelization of countries which had already yielded their nominal adherence to the Church at a previous epoch.

Columban seems, like others of his class and calling, to have been of a restless nature. He became wearied of the hardships or difficulties of his position on the frontier of German paganism, and at a later period migrated to the further side of the Alps, where he preached among the relapsed believers of Northern Italy, and founded the illustrious monastery of Bobbio. He left behind him, however, a faithful band of Irish auxiliaries, one of whom, known as St. Gall, was planted in the monastery which has perpetuated his name, near the Lake of Constance. But the energies of these Irish missionaries seem to have been devoted less to the propagation of the faith among the heathen, than to the maintenance of their own special tenets or usages in matters on which they differed from the Southern Church and the authority of the Roman see. These tenets were apparently simpler, and in some respects of purer character, than those with which they were confronted.

The Irish strenuously asserted the right of the clergy to marry; but there was no point on which they were so unyielding as the observance of Easter according to the Eastern computation, which is held to have been originally the Roman also. The influence of the popes, or rather the spirit of the time, proved too strong for them on this and all other points, and they ultimately fell under the monastic rule of St. Benedict, to which we may fairly allow that their principal institutions owe the reputation for industry and learning which they so long held and deserved.

Meanwhile, the Gospel was carried by other Irish adventurers further into the interior of Germany. Thuringia, the region between the Main and the Danube, had enjoyed but little of the Roman civilization, and hitherto still less of Christian preaching. But at the end of the sixth century a queen, named Radegunde, had abjured the errors of her pagan ancestors, and had manifested her zeal as a convert in burning the temple of the national idols. Her people, however, still continued to be pagan, and when the Austrasian king Dagobert visited the country, which was nominally subject to him, in 622, he found it still deeply sunk in heathen barbarianism. Odilon, one of the chiefs, having a sick relative at his house when he was about to join the sovereign's retinue, gave orders to cut off the dying man's head and burn the corpse; for such was the law of his religion, that those who were in mortal sickness should be slain with the sword, in order to entitle them to enter the gates of Odin's Valhalla. Still later, an Irish bishop, named Kilian, accompanied by the priest Colman and the deacon Totnan, penetrated as far as Wurzburg, on the Main, and having satisfied himself that the spot was suitable for a Christian mission, repaired in person to Rome, to engage the pope Conon to allow him to evangelize the still pagan Thuringians. The prince of the country solicited baptism; but, as he had for his queen the wife or possibly the widow of his brother, and the Church demanded the abandonment of this incestuous union, the wicked woman, like another Herodias, procured his assassination, together with that of his companions. The story, whether true or not, may be taken as a type of the contest which the Church undoubtedly waged against the corrupt passions of the kings and queens of paganism. There seems to have been no lack of courage among the preachers of the faith, in the attacks they made upon wickedness in high places, as long at least as they were sustained by the ardour of missionary zeal. It was rather when the faith became generally accepted and established, that its preachers were themselves corrupted, and tempted to connive at the sin of great men of the world, who had become only nominally their disciples.

The Irish missions, headed, as we have seen, by Columban, form the most interesting feature in the progress of the faith in Germany through the period which has now been reviewed. The seventh century saw another succession of Christian preachers engaged in the countries south of the Danube, at which we can only glance as we pass on. Gregory the Great had devoted himself zealously to the recovery of that country, which had been already evangelized, from the Arian to the orthodox side. Clotaire II., of Austrasia, sent two monks from Luxeuil, named Eustatius and Agilus, to preach in Bavaria, but heterodoxy or paganism still made an obstinate resistance. When Bishop Emmeran, of Poitiers, arrived at Ratisbon, in the middle of the seventh century, he was astonished at the remains still existing there of the old Roman grandeur, a city enclosed in walls, with a palace, and many churches; but the people were sunk in superstitions, and partook, as he complained, on the same day and from the same chalice of the blood of Christ and the wine offered to demons. Though supported by the influence of the reigning duke, he fell a martyr at the hands of these barbarous novices. At the close of the century his place was supplied by Rupert, Bishop of Worms, who baptized the prince of his own time, together with a multitude of his chiefs and warriors. After wandering far along the course of the Danube he turned aside to seek out the remains of the Roman city of Juvava, or Salzburg, once a flourishing colony, founded by Hadrian, the imposing remains of which sheltered now a mere handful of simple mountaineers. Here Rupert erected a church and a monastery, and persuaded the rude people to gather round them and found a new city. Thence returning to France, he sought out and brought back with him a troop of monks and nuns, according to the fashion then prevailing, that the sexes might glorify the laws of Christian purity by dwelling closely together, and preserve their chastity for the admiration of the heathen. The legend of Rupert and his abbess Ehrentrude affords a pleasing specimen of the manners of the time. The saint received a revelation of his approaching departure. Addressing his spiritual sister, "I desire," he said, "to speak with you in private. God has just made known to me that He is about to take me; let me, then, beg your prayers for my soul." The virgin burst into tears, and answered, "O sir, if you are going to die, better it were that I should die before you." The bishop replied, "Suffer not yourself, my dear sister, to wish to leave the world before the time, for that were a great sin." Then Ehrentrude threw herself at the bishop's feet: "My lord and father," she cried, "remember that you caused me to abandon my country, and now you will leave me alone and an orphan. I beg of you one thing only, which is this, that if I cannot go before you, I may, at least, through your intercession, obtain the boon of closely following you." This Rupert promised, and so, after conversing together about the life to come, they took their last farewell of each other sorrowfully. On the day of the Lord's resurrection, Rupert, having celebrated, and blessed the people, knelt in prayer, and so passed away. Soon after Ehrentrude, having prayed much for the soul of her friend and brother, heard at night a voice calling her, fell sick and departed to the Lord.

Towards the middle of the eighth century three pilgrims from Ireland presented themselves to the Bavarians,—the hermit Alto, whose cell became the cradle of the abbey of Altenmunster; the bishop Dobla, surnamed the Grecian, perhaps for his reputed knowledge of the ancient tongue of Christendom; and the monk Virgilius, who built a church at

Salzburg, which he dedicated to St. Rupert, and in which he set up his throne. Mention is made of his labours among the Carinthians; but a peculiar interest may attach to him from the circumstance of his having guessed apparently the existence of antipodes on the surface of the earth's globe, for which, it is said, that he was denounced to the Holy See and condemned. The advocates of the Papacy repudiate this allegation against Rome; they admit, however, that the holy missionary, St. Boniface, of whom we are to hear more presently, took an unfair prejudice against this good man, and did actually accuse him to Pope Zachary of various errors, and particularly of having declared that there existed beneath the earth another world, and that there dwelt upon it a race of men who had no share in the sin of Adam or in redemption through Christ. Doubtless the pontiff inquired into the truth of this charge, but as we find that the accused was raised soon afterwards to the archbishopric of Salzburg, and at a later period was canonized by a successor of Zachary, we are bound, it is said, to conclude that he found means of reconciling the conjecture, which was not a new one, with the dogmas of the Catholic faith. And so, in the progress of discovery, first north and west, and afterwards south and east, the fact that this earth is globular became gradually recognised and accepted by the Church.[1]

But if we take the year 700 as the limit to which we have now advanced, we may observe that at that

[1] Ozanam, 'Études Germaniques,' ii. 134.

period the conversion of the faith in Germany had extended, speaking generally, to the Franks, the Allemanni, and the Bavarians, to whom may be added the Burgundions settled in France, the Visigoths in Spain and Italy, the Ostrogoths in the valley of the Lower Danube. The German nations were beginning to settle down into the type of civilization which they had either developed themselves, or had received from the Latin races of the south. The peoples throughout central Europe were becoming fixed in their habitations, marking out the limits of their several communities, building themselves cities, laying out lines of trade, establishing depositories of their respective products, and protecting their social life with definite codes of law. In the legislation of Thierry, of Childebert, Clotaire, and Dagobert, we may still trace, it is said, the pagan customs on which it is primarily founded; but these are gradually subjected to the more humane and beneficent principles of the canon and the civil law. The rights of property, secured first of all to the Church, are placed generally under the sanction of the religious sentiment. Human life, however rudely its value may be estimated by a pecuniary composition, rising by successive stages from the slave to the clerk and to the bishop, is at least sheltered thereby from the caprice and violence of passion. The right of sanctuary afforded to the fierce crimes of a fierce and impetuous people, secures an interval for the operation of mercy and legal justice. On all these points it may be truly urged that the Christian Church, while legislating immediately for her own interests, was effectually working in the interest of the people of whom she had claimed to take spiritual charge. In all the countries at which we have now glanced the Church had replaced the Empire, the Christian civilization was gradually replacing the Roman; the popes occupied, in fact, the throne of Augustus and Hadrian. We now go further afield in the footsteps of the Christian missionaries, who are about to invade new territories beyond the Rhine and the Danube, from which Augustus and Hadrian had turned back in dismay.

CHAPTER VII.

ST. BONIFACE THE APOSTLE OF WESTERN GERMANY.

THE Irish missionaries were succeeded by the English. From the end of the sixth century the shores of our island had been sought, after a long interval, by preachers of Christianity sent direct from Rome, and pledged to implicit obedience to the tenets of the Latin Church. The Anglo-Saxon invasion had already almost wholly eradicated the seeds of Christian faith which had been previously fostered by the British Church; it was only among the mountains of Wales, or on its western coast, that the remnant of the Celtic population still retained its earlier knowledge of the Gospel. St. Augustine undertook the double work of converting the pagans to Christianity, and of bringing over the British believers to the Roman obedience. In both these undertakings he proved eminently successful. King Ethelbert, of Kent, and his pious queen, Bertha, supported him from the moment of his first appearance before them. The recent conquerors of Britain, speedily converted by his teaching, became the foremost and most devoted adherents of the Holy See. The piety of their kings and princes, of their bishops and monks, acquired celebrity throughout Christendom. In England, even more eminently than elsewhere, queens showed themselves nursing mothers of the Church of Christ. Many of these pious women, such as Etheldreda and Ermenilda at Ely, Werburga at Chester, and Hilda at Whitby, attained the honour of canonization. As time went on, when the faith had become fully established throughout their country, the most ardent spirits of the English Church began to turn their views towards the evangelization of the heathen abroad. The foremost, and also the most illustrious of this band of missionaries, was Winfrid, who, under the assumed name of Bonifacius, has obtained the title of the Apostle of Germany.

In the year 718, this devoted aspirant to missionary honours presented himself to the Pope, Gregory II., at Rome, and requested employment under his sanction, bringing with him a recommendation from Daniel, the Bishop of Winchester. Winfrid was at the time about forty years old. He was born at Crediton, in Devonshire, and had studied or meditated as a monk in a monastery at Exeter, and also at a place called Nutscell, on the banks of the Solent. Impatient, it would seem, of conventual restraints, he had crossed the sea two years earlier, and made the experiment of preaching among the Frisian heathens. His course had been arrested by the war waged between Charles Martel, the Austrasian mayor of the palace, and this people, under their chief Radbod. The seeds of the faith so lately sown seemed to have been scattered, and Winfrid had returned, much dejected, to England. But he retired only to make a more auspicious advance soon afterwards. The Pope, having made a close examination of his faith, gave him the sanction he required for carrying the Gospel among the heathen, wherever and as far as he should choose. Winfrid recrossed the Alps, bearing with him credentials which secured him the respect of all Christian princes and ministers, and which might invest him with superior sanctity even in the eyes of heretics and half-believers. He directed his course through Lombardy, Bavaria, Thuringia, and Eastern France, for even in these countries he might still find multitudes not yet subjected to the Roman obedience, which, no doubt, he inculcated, at the same time that he enforced the broader rites and doctrines of Christianity itself. As soon, however, as the movement of the Frisians had been composed on the death of Radbod, he made their country, as before, his immediate object. It seems that he found there another Englishman named Willibrord exercising the functions of a bishop among the new converts. To this leader he attached himself, and at the end of three years was asked by him to receive episcopal consecration, and succeed to a charge which had become too heavy for the older prelate. But Winfrid, or Boniface, as he seems already to have called himself, betokening, perhaps, his devotion to works of practical piety, shrank from such responsibility, and even fled from the country, determined to bury himself in regions which seemed to him still to require the restoration of their faith, after the many shocks to which constant war and devastation had subjected it. Such a region was Thuringia, a country then still abounding in heathens, where, indeed, the faith once preached by St. Kilian and others, had been shaken off by multitudes in a later generation. Here he converted some, and recovered others. Here he obtained grants of land, and founded monasteries. But his restless spirit soon impelled him to go further afield. Among the Hessians and the Saxons he baptized, it is said, many thousands, and after sending a faithful account of his doings to the pope at Rome, whose devoted servant he esteemed himself, he proceeded in person to make his profession of faith at the Vatican, and prove that his teaching had been fully in accord with the instructions which had been laid upon him. Pope Gregory was well satisfied, and in the year 725 he insisted on consecrating his zealous minister to be what was called a regionary, that is, a missionary bishop without local jurisdiction; but it was to the country of the Thuringians and the Saxons that Boniface was specially directed. He was placed, moreover, under the protection of Charles Martel, with whom the Roman pontiff had already exchanged compliments and services, with a view to the closer relations between the Holy See and the court of France which followed in the next generation. The pious and simple-minded Boniface was soon shocked at the corruption of manners at the palace; but he turned his thoughts to the work to which he was more particularly summoned, and contented himself with seeking counsel from his early patron Daniel as to how he should conduct himself in his spiritual warfare with the German pagans. The reply of Daniel is reported by the biographers, and may serve as an indication of the kind of preaching which was deemed most edifying and most suitable to the promotion of the truth in Christ.

"You must not," said Daniel, "raise your voice against the genealogies of their false divinities. Rather let them declare to you how their gods were born one from another by carnal copulation; then you can readily show that gods and goddesses of this human origin can be no other than human beings, and that as they have once begun to exist, they cannot continue to exist for ever. Thence proceed to ask them, whether the world has had a commencement, or whether it is eternal, and if it has commenced, who has created it? Again, ask them, where did these deities, who have been born, reside before the creation of the world? If they say that the world is eternal, who was it that governed it before the birth of the gods? How did they bring the world under subjection to their laws, seeing that the world had no need of them? Whence came the firstborn among themselves, and by whom was he generated from whom all the rest are descended? And further, ask them, whether they think the gods ought to be honoured for the sake of temporal and present happiness, or of the future and eternal? If they say, for temporal happiness, then let them show in what way are the pagans better off than the Christians. You shall address them with these and suchlike objections, not by way of provocation and insult, but with the greatest moderation and mildness. And from time to time you shall compare their superstitions with the Christian dogmas, touching them lightly indeed, so that the pagans may remain confounded rather than exasperated, that they may blush at the absurdity of their prejudices, and not suppose that we are ignorant of their false opinions and sinful practices. Further, you shall present to them the greatness of the Christian world, compared with which they are themselves so insignificant. And to prevent their boasting the immemorial sovereignty of their idols, take heed to teach them that idols were indeed adored through the whole world until the time when the world was reconciled to God by the grace of Jesus Christ."[1]

Such counsels, it might be hoped, would gradually undermine the superstitious usages of the heathens, which it would be indiscreet to attack directly. Seconded, as they doubtless were, by the austere lives and kindly deeds of the preachers, they could not fail to make a happy impression. Nevertheless, while many of the natives made public abjuration of their false religion, a remnant still continued to sacrifice, some openly, others more or less in secret, to the trees or fountains which they had learnt to regard as holy, and to practise, as of old, the incantations and sorceries of their ancestors. Then it was that the holy apostle determined to aim a bolder blow at the root of the pagan idolatry by attacking a tree of marvellous height and size which the people regarded with special veneration, under the name of the oak of Thor. A rude multitude rushed to defend, with their arms, this monument of the faith of their fathers, and to

[1] Ozanam, 'Études Germaniques,' ii. 180, from the 'Epistolæ Bonifacii.'

slay the impious assailant of their gods. The bishop appeared, surrounded by his little band of clergy. He himself levelled the first blows of his axe at the mighty trunk, and before the pagans had recovered from their amazement, a vehement blast of wind came, as a sign from heaven, and overthrew it, breaking it at the same time into four pieces of equal length. The spirit of the angry crowd was completely mastered, and they at once acknowledged the power of the God of the Christians.[1]

[1] It may be interesting to compare the well-known description as given by Gibbon, of the overthrow of Serapis, at Alexandria, by the order of the emperor Theodosius :—

"The colossal statue of Serapis was involved in the ruin of his temple and religion......It was confidently affirmed that, if any impious hand should dare to violate the majesty of God, the heavens and the earth would instantly return to their original chaos. An intrepid soldier, animated by zeal, and armed with a weighty battle-axe, ascended the ladder; and even the Christian multitude expected, with some anxiety, the event of the combat. He aimed a vigorous stroke against the cheek of Serapis; the cheek fell to the ground: the thunder was still silent, and both the heavens and the earth continued to preserve their accustomed order and tranquillity. The victorious soldier repeated his blow: the huge idol was overthrown and broken in pieces; and the limbs of Serapis were ignominiously dragged through the streets of Alexandria. His mangled carcase was burnt in the Amphitheatre amidst the shouts of the populace, and many persons attributed their conversion to this discovery of the impotence of their tutelar deity......After the fall of Serapis some hopes were still entertained by the pagans that the Nile would refuse his annual supply to the impious masters of Egypt; and the extraordinary delay of the inundation seemed to announce the displeasure of the river-god. But this delay was soon compensated by the rapid swell of the waters," &c.

The well-known veneration of the Germans for the giants of

This notable success was followed by a rapid multiplication of converts, and Boniface continued to follow it up with singular energy and perseverance. The construction of an oratory with the wood of the famous tree was the commencement of an era of ecclesiastical building. Churches were founded in many places, such as Altenberg, Ohrdruff, Frizlar, and Erfurt, on the borders of Thuringia, and in the heart of Germany. The next thing was to secure a succession of priests and monks to turn them to spiritual account.

It was to England rather than to France that the apostle now resorted. He addressed himself to the kings, the bishops, and the holy women who ruled in so many monasteries of the isle of saints. He asked for devoted missionaries first of all; but he demanded also the most necessary articles of ecclesiastical furniture, the priestly robes, ornaments, bells, and more especially, books. In his letters, a number of which have been preserved for us, he specifies the "questions" of St. Augustine of Canterbury together with the "replies" of Gregory the Great, the "Acts, Sufferings, Passions of the Martyrs, the "Commentaries of the fathers on St. Paul," a volume of the prophets, written plainly, without contractions, for the

their forests had been long before marked. Lucan ascribes a similar sentiment even to the Roman legionaries, when commanded by Cæsar to demolish the sacred grove of the Druids :—

"Sed fortes tremuere manus, motique ruendâ
Religione loci, in robora sacra ferirent
In sua acredebant redituras membra secures."
Pharsal. iii.

sake of his own weak eyesight. He implored the abbess Eadburga to have the epistles of St. Paul transcribed for him in gilt letters, "in order to give the Holy Scripture due honour in the eyes of the carnal men," or pagans. These appeals were, no doubt, fully responded to. A multitude of missionaries went forth to him from the convents of England, placed themselves under his orders, and became famous in turn as devoted preachers of the faith among the Germans. Such were Willibald, Wunnibald, Witta, Wigbert, to whom may be added Sturm, a native of Bavaria, who became, under the direction of Boniface, the founder of the illustrious abbey of Fulda. To these were added a number of female devotees, some widows, some virgins, who came forth also from their quiet convents to encounter the perils and hardships of labour among the heathens, such as Chunigild and Chunigrat, Thecla and Lioba, some celebrated for their beauty, some for their learning, all for their piety and the happy influence they exerted upon the chivalrous character even of the wild Thuringians.

The teaching of Boniface, as well as of other eminent leaders in the great cause of conversion, was of two kinds, which it may be well to keep in distinction: to the priests and monks under his direction he constantly addressed himself as lieutenant of the Pope, and champion of the Roman Church. With them his contention was against every inclination to heretical opinions and wilful practices. He sternly put down all personal independence. He aimed particularly at the suppression of the views of the Irish Church, such as had been preached by

Columban, and required from all his ministers the most entire obedience to the one mother of all the Churches. The unbounded influence he acquired over his assistants in the holy work before him secured him complete success in this part of the mission he had imposed upon himself. There is no doubt that through the greater part of his career he effectually checked the opinions of which the Church of Rome was most jealous. Nor need we hesitate to admit that in the crisis of the barbarian conversion no greater service could be done to the general cause of Christianity than by the course which he thus pursued, in requiring unity of doctrine and uniformity of practice wherever his influence extended. We find, however, that in addressing himself to the simple people among whom he had undertaken to diffuse the pure light of revealed truth, he confined his teaching, when this principle was once established, to the plainest doctrines of Scripture, and the fundamental rules of universal morality. A collection of his homilies, to the number of fifteen, has been preserved by his followers and admirers, some of which treat directly of the cardinal dogmas of the Incarnation, the Nativity, and the Resurrection, but in all of them the chief points of faith and worship, of morality and discipline, find their proper place. In the fifteenth, for instance, of the series he thus instructs the neophytes whom he has just introduced into the Church by baptism:—

"Hearken, my brethren, and consider with attention what it is that you have now renounced. You have renounced the devil, his works and his vanities.

What are the works of the devil? Pride, idolatry, envy, murder, calumny, falsehood, perjury, hatred, fornication, adultery, and everything that defiles a man; such as stealing, bearing false witness, gluttony, drunkenness, strife, and evil-speaking. Devotion to sorceries and incantations, belief in witches and man-wolves, wearing of amulets, rebellion against God; these, and such as these, are works of the devil. These you have renounced at your baptism, and, as says the apostle, they who do these things shall not enter into the kingdom of heaven. But, believing as we do, that you, by God's mercy, have abandoned all these iniquities, both in thought and deed, it remains to remind you, my well-beloved brethren, of what you have promised in your baptism to do in their stead. For, first of all, you have promised to believe in God Almighty, in His Son Jesus Christ, and in the Holy Ghost, one only God in a perfect Trinity. See what are the commandments you must keep. You must love God, whom you have confessed, with all your heart, with all your soul, with all your strength; and next, your neighbour as yourself; be patient, merciful, good, and chaste; teach your children the fear of God, and teach your servants also; make peace where there are quarrels; let him who is a judge refuse to accept gifts, for gifts blind the judgment even of the wise. Remember to observe the Lord's day, and betake yourselves to the church to pray there not to amuse yourselves with empty babbling. Give alms according to your ability, practise hospitality, visit the sick, minister to the necessities of widows and of orphans, pay your tithes to the Church;

do nought to any that you would not have done to you; fear none but God, but fear Him always; believe in the coming of Christ, in the resurrection of the flesh, and in the universal judgment."

It will be observed that there is hardly an expression in this discourse which does not correspond exactly with the catechetical teaching of the modern Church, the same which we have inherited from the earliest antiquity. It contains what all Christians must admit as the very ground and basis of their own faith. But besides this it indicates a Church established and organized on the primitive lines, with its system of common prayer, common preaching, and common discipline. It seems to show that the Thuringian Church, as we may call it, which is said to have comprised not fewer than a hundred thousand of the apostle's own converts, had struck its roots in the soil of Germany, and become a spiritual power of real importance in Christendom. Nevertheless, the position of this Church, so seemingly prosperous, was still insecure. The convulsions which had already so often overturned the establishment of the Christian missionaries beyond the Rhine, and revived the false teaching of the heretics, and even the worst form of the heathen idolatries, were not yet exhausted. The rude, half-reclaimed multitude rebelled against the austerity of the Christian discipline. They resented the restraints it put upon them, especially in regard to the relations of the sexes. It would seem that the priests themselves murmured against the rules thus imposed upon them. Boniface was disposed to yield rather than behold his good work undone before his eyes; but he laid the case before the Pope, Gregory II., and the commands of Rome were firm and inflexible. He determined to persevere, and as a reward and an encouragement he received from the pope the pallium, which conferred upon him the dignity of a metropolitan, and gave him spiritual authority over the various bishoprics which were now founded in Western Germany. From his archiepiscopal seat at Mayence he superintended the sees of Cologne, Tongres, Worms, Spires, and Utrecht.

But the Church in Western Germany was soon subjected to a severer trial than the passing discontents which had for a moment shaken the nerves of the dismayed apostle. Charles Martel, the warlike chief of the Frankish sovereignty, who had acted as a pioneer of the Church in quelling the barbarian races of Friesland and other northern districts, and preparing the soil for the seeds of Christianity which the missionaries carried after him, proved but a worldly champion of the faith of Christ. He had succeeded in uniting the two Frankish kingdoms of Austrasia and Neustria; he had raised himself from the position of Mayor of the Palace to the royal dignity, and the pope (who looked to him as his surest support against both the Lombards and the Greek emperors) had confirmed his usurpation. But his power had been sorely tried by the irruption of the Saracens into the south of France, which they threatened to sever completely from his dominions. The fate of Western Christendom trembled in the balance, for on the fall of the Christian kingdom of the Franks there would remain no other power beyond the Adriatic which could hope to maintain itself against the conquering fanatics of Arabia. Charles Martel had put forth all his energies, and his efforts had been crowned with the great victory which is not perhaps unjustly deemed to constitute the most important crisis in modern European history. The danger had been averted, and the Saracen invasion was finally checked. But the effort had cost much in many ways. It seems that the conqueror, worldly as he was, had not scrupled to surrender the Church, its dignities, and its possessions, to the worldly powers which had been long hovering around it, jealous of its influence, greedy of its substance, hostile to the spiritual authority which it wielded over the consciences of the people. The golden age of Christian purity and of Latin orthodoxy in France and of Western Germany had been of short duration. We now hear of the rapid degeneracy which overtook it in the corruption of its priests and bishops, the corruption of its faith, the relaxation of its morals, the revival of the heresies of an earlier age, and the imminent restoration of paganism.

The subtle heresies of the Greeks, it was said, were imported into Western Germany through the agency of the Goths and the Heruli. Two monks from Africa are reported to have introduced the doctrines of the Manichees; there were bishops without sees, priests without a charge or mission, clerks who lived in concubinage and drunkenness. Particular mention is made of one Irishman with the Latin name of Clement, and another with the Teutonic name of Adalbert, of whom, the first decried the authority of the Church and the fathers; the second boasted that he had visions of angels, and was a worker of miracles. How far these statements are genuine, and how far they indicate merely a rebellion against the tyrannical domination of the Roman see, may be a question. We may allow, however, that, under the circumstances of the age, it was only through obedience to the Roman discipline that the faith could, humanly speaking, have maintained itself successfully against the evil passions of human nature; and the instinct was not wholly at fault which at such a crisis regarded a falling away from Rome as a revolt against Christianity itself.

In the face of the double danger which then threatened the Church in Germany, Boniface undertook another pilgrimage to Rome, and threw himself once more at the feet of the pontiff who now occupied the seat, the third Gregory. This was in the year 738. He was received, it seems, with great distinction, as might be expected, by the pope and his clergy, and not less, it is said, by the whole multitude of pilgrims from the West—Franks, Bavarians, and Anglo-Saxons. He spent a whole year at Rome in completing his arrangements for the extension and government of the Church in Germany, and in paying his devotion at the tombs of the saints, whose prayers he solicited for the success of his great mission. On leaving the holy city he was specially directed to betake himself to the Allemanni and the Bavarians, and to establish or revive the seats of the bishops on the banks of the Danube. Under the patronage of the secular power he convened a synod, which decreed the division of his allotted province among the four sees of Salzburg, Freisingen, Ratisbon, and Passau. Heresy and superstition quickly disappeared, the work of Severinus, of Rupert, and of Virgil rose from its ruins; and after some interval Boniface found himself at liberty to proceed, as he wished, northward. On the death of Charles Martel, in 742, his son and successor Carloman, who was a devoted servant of the Church, allowed another synod to meet, and sanction the establishment of three new sees at Wurzburg, Bamberg, and Eichstadt, in Franconia, under Boniface, as their metropolitan. The Churches throughout the country were restored to their rights and possessions; schismatical and irregular priests were degraded, and the strict discipline of the Latin Church revived and enforced upon both clergy and laity. Still there remained many vestiges of ancient paganism to be rooted out; such as the observation of auguries and omens, the lighting of fires in honour of false gods, and sacrifices at funerals. At a third synod held at Leptines near Cambrai, in the following year, the inmates of the cloisters, monks and abbots, submitted to the monastic rule of St. Benedict. Among many rules of holy life and practice which were also decreed, was one by which the selling of Christian captives to the heathen was specifically forbidden, a tardy and trifling step towards the recognition of the sin of man keeping man in slavery at all. It must be added that Carloman, devout as he was, did not refuse to accept, in consideration of the weight of temporal power which it threw into the scale, a portion of the ecclesiastical property of which he had saved the bulk for the Church. But having effected this arrangement for the benefit of his successors, he quickly descended from his throne, and buried himself in a monastery. The advantages thus secured to the Church extended only to Austrasia. In the following year, Pepin, with the sanction of a council held at Soissons, conferred the same benefits upon the kingdom of Neustria, which soon became reunited with the eastern portion of the great Frankish dominion. The compact which was now formed between the sovereigns of France and their clergy has lasted more or less strictly ever since. It was solemnized once for all on the notable occasion, when, in the year 752, the Frankish warriors elevated Pepin on their bucklers at Soissons, and the bishops gave him the holy unction of the kings of Israel. This rite, now first introduced among the Franks, was borrowed, it is said, from the liturgy of the Anglo-Saxons; and some of the chronicles do not fail to assert accordingly, that it was Boniface who anointed Pepin king of France.[1] This fact, indeed, is still subject to debate. The French critics are unwilling to surrender to an Englishman the glory of having consecrated the kingdom of Clovis and of St. Louis, as the eldest daughter of the Church.

Among the most permanent of the apostle's works was the institution of the abbey of Fulda, which retained through the middle ages a position equal in renown to that of St. Gall in Switzerland. It took its name indeed, not from the holy man, Sturm, who was deputed by Boniface to lay its foundations, but from the spot on the banks of the little river so called, which

---

[1] Ozanam, 'Études Germaniques,' ii. 102.

falls into the upper waters of the Weser, in the midst of what was known in those days as the great Thuringian forest. It was in the year 751 that Boniface solicited of the Holy See a privilege which should place the new institution, for which he entertained a special affection, beyond episcopal jurisdiction. It was, perhaps, the last infirmity of a noble mind which inspired him with the idea that a foundation of his own, subjected by himself to the rules of the conventual order in which he reposed implicit confidence, would be more exempt from the chances and changes of the world around it than if it were placed under the direct government of the bishops, so many of whom he had found, by his own experience, unfaithful to their high calling. It may be worth while to recite the words with which he preferred his petition to the Holy See. "There is a wild spot," he said, "in the depths of a vast solitude, in the midst of the people over whom my apostleship extends, where I have raised a monastery for brethren under the rule of St. Benedict, men bound to severe abstinence, forbidden the use of wine or meat or of domestic service, who shall be content with the labour of their own hands. I have acquired this possession from divers pious persons, and especially from Carloman, prince of the Franks, and I have dedicated it in the name of the Saviour. There it is, that with the good-will of your Holiness, I have determined to give repose for a few days to my body, broken as it is by old age, and to choose a place of sepulture; for the spot is in the neighbourhood of the four nations, to which, by the grace of God, I have proclaimed the word of Christ." The nations, we may suppose, were the Franks, the Thuringians, the Bavarians, and the Allemanni. The privilege was accorded, and "hence commenced the greatness of the powerful abbey of Fulda, which, along with the kindred foundation of Saint Gall, realized eventually the ideal of the monastic colonies of England, and carried into central Germany all the lights of the island of saints."[1]

We might have expected, perhaps, that Boniface, so deeply interested, as he expressed himself, in the fortunes of his beloved monastery, would have carried out the design he here intimates, and spent the remainder of his days in repose and devotion. But not such was his nature. Hardly had he sighed out these pious aspirations when he suddenly renounced, not only his vision of quiet solitude at Fulda, but the actual duties of his appointed station as the metropolitan of an episcopal province, and rushed off from the sphere of organized government before him to commence a missionary incursion into the barbarous regions of northern Germany, from which, in his more vigorous years, he had been repulsed. His tender nature was pained to hear of the falling away of the converts whom he had formerly made, or fancied he had made, in those heathen regions. He imagined that the call he had recognised in his youth was again repeated. He must go forth once more, and begin over again the labours of his earliest mission. At the age of seventy-five he fondly thought that he could effect what he had failed in effecting

---

[1] Ozanam, ii. 205, who cites his authorities from the collections of Mabillon, Pertz, &c.

forty years earlier. He transferred to his disciple Lull the archiepiscopal dignity, no doubt with the necessary sanction, giving him in charge to complete the fabric of his monastery at Fulda, as well as to watch over the faith and morals of the disciples throughout Western Germany. "For myself," he added, "I must start betimes, for the day of my departure is at hand. For this final departure I have long wished; get everything ready for me, and particularly take heed to place in the chest which holds my books the shroud in which my body shall presently be wrapped." Thereupon he summoned to his side Eoban the bishop, Walther and Wintrig priests, Hamund, Skirbald, and Bosa deacons, Waccar, Gundwaccar, Illesher, and Bathowuld monks, and so the whole party descended together the stream of the Rhine as far as Utrecht, and speedily commenced the work of preaching the Gospel to the natives. Many thousands, men, women, and children, received baptism at their hands.

But this first success proved illusory, and disaster was at hand. Before the middle of the same year, when the great missionary had made preparation for the celebration of the Christian sacraments on an imposing scale at a place now called Dockum, on the northern coast of Friesland—so far had he penetrated on his perilous enterprise—and was surrounded by a multitude of peaceful candidates for baptism, he was suddenly attacked by a host of savage heathens fully armed for his destruction. His attendants would have made some attempt at defence, but their master himself forbade them. Encircled by his priests, and bearing in his hands the sacred relics which he always carried with him in his missions, Boniface exclaimed, "Cease, my children, from strife. Know ye not that the holy Scripture commands us to return good for evil? This is the day which I have long desired, and the hour of our deliverance is at hand. Be strong in the Lord; put your hope in Him, and He will save your souls." With these and other like pious words he comforted and sustained his companions, till they all died the death of martyrs together under the sword of the persecutor. It was fondly recorded, that when their murderers rushed into their victims' tents in search for gold and silver, they found in them nothing but holy books and relics, and the wine reserved for the eucharistic sacrifice. Disappointed and mortified they quarrelled and fell upon one another, till the Christian party, still hovering round, attacked them in turn and utterly destroyed them. Nevertheless, however signal this speedy retribution might be, it would seem that the last enterprise of Boniface perished with him. His brief incursion into the country of the Frisians left apparently no permanent impression. His was the last attempt to carry the blessed truth of the Gospel by peaceful missions into the territory of the northern Germans. The sword, it might seem, alone remained as an instrument for the propagation of the faith; and the time was ripe for at last resorting to the literal interpretation of the text—so often and so fatally abused—"Compel them to come in."

CHAPTER VIII.

CHARLEMAGNE AND THE FORCIBLE CONVERSION OF THE SAXONS.

EIGHT hundred years before the time at which we have now arrived, the central regions of Europe were divided between two powers, which stood in a state of constant hostility to one another, and disputed with alternate success the empire of the world. The Germans and the Romans were for the most part separated by the broad streams of the Rhine or the Danube. The Germans had, in earlier times, made many incursions into the territories beyond these frontier lines, but from time to time they had been beaten back to them again, and driven still further into the interior. Through the reign of Augustus the struggle had been pertinaciously continued. After a century of contest the limits of the two powers seemed to be more definitely settled, and for a second and a third century no great change had been made in the limits to which they were respectively confined. The language, the manners, the polity, the religion of the two nations had been different from the first; and no doubt the enmity between them had been partly fostered by their inability to understand or appreciate one another. The Romans had spoken with great horror of the barbarous superstitions of the Germans, unaware, it would seem, that there was hardly less barbarity in their own. The Germans had retorted with indignant disgust at the vices and effeminacy of their more polished adversaries. Nevertheless, the contest between the two races was never carried on in the name of religion or morality. It would be difficult, perhaps, to specify any war between the nations of ancient history which professed to be undertaken on account of a moral or religious sentiment. The wars of Charlemagne, the king of the Franks, the emperor, as he was at last designated, of the Romans, against the Saxon pagans, may deserve to be entitled the first crusade, the first contest avowedly undertaken for the overthrow of the false and the propagation of the true religion. Whatever worldly motives may have mingled in the mind of an ambitious ruler, such as Charlemagne, the pretext he put forth was distinctly a religious one, the passions to which he appealed were those which too often disguise themselves under the name of religion. The establishment of a Christian Church and an ecclesiastical hierarchy throughout the realms he conquered from heathendom was the end he sought, and this end he actually attained.

The pressure of a powerful and well-organized military state like that of the Austrasian Franks, had put a severe constraint upon the restless passions of the northern barbarians. Now, again, as so many times before, they chafed at their confinement within their hereditary limits, and longed to roam freely over the continent as their fathers had so often done before them. It may be that the ostentatious humility, and the peaceable character of the Christian Churches and the missionaries with whom they had come in contact on their borders, had nurtured in them some contempt for the ruling powers of the empire which they were impatient to invade. Proud of their long hair, which they regarded as a sign of bodily strength, and honoured as the distinguishing mark of their illustrious race, they looked with contempt on the smooth chins and shaven crowns of the monks and priests who pretended to correct their laws and manners. The Christian preachers, on the other hand, were shocked at the brutal habits of the people whom they strove to win to a purer life and doctrine; they found among them too evident proofs, not only of cruel and sanguinary rites, but of human sacrifices, and even of cannibalism; and the accounts they gave of the horrors they had heard or witnessed excited the devotees of their own country to a pitch of exterminating zeal. Let the Saxons be converted, they cried, or let them be massacred; be it either one or the other, it mattered, perhaps, little which.

In the eighth century the Saxons, originally confined to the Cimbric Chersonesus, had become a wide-spread and an established power in Northern or Lower Germany. The limits of their dominion were at that time pretty clearly defined; namely, from the Elbe to the Yssel, including the three districts of Estphalia, Westphalia, and Angaria, which lay between them. These districts embraced, in fact, the modern Hanover and Oldenburg, together with a small portion of Prussia. The country was for the most part flat; but the forests with which it was covered impeded communication, and checked any progress in civilization. There seem to be no traces of cities among the Saxons, nor, of course, of municipal institutions, few perhaps of regular cultivation. The people remained, like the ancient Germans in the time of Tacitus, immersed in almost primitive barbarism, while, with a teeming and constantly overflowing population to maintain, they were repeatedly impelled to make inroads upon the more fruitful regions beyond their borders. No doubt the original quarrel between the Franks and the Saxons was kept alive by the pressure of this vital necessity; but, besides this, a further, and perhaps a not less bitter, enmity was engendered between them by the mutual repugnance of their respective creeds. The Franks of Germany may have been even yet but half converted to the faith of the Gospel, much less than half converted to its precepts; but they had acquired just so much concern or interest in it as made them keen to mark the defects of their neighbours' theology, and fierce in resenting it as an affront to themselves. The Saxons were still at this period wholly heathens. No Christian mission, it would seem, had hitherto penetrated into their fastnesses. They repudiated all connection with the South, and boasted of their descent from the hordes of pirates who had infested for centuries the coasts of Scandinavia. The legend that their ancestors had been among the companions of Alexander in his Asiatic conquests, was no doubt the invention of the stranger, in which they took, at least till a later period, no national interest. But it was more or less skilfully connected with the popular tradition, according to which the Saxons derived their name from the sahs or knives which they drew on a famous occasion to cut the throats of the Thuringian chiefs who had invited them to a peaceful conference.

Hitherto the Saxons had been chiefly known as wanderers on the ocean, who had for ages infested the shores of Gaul and Britain, and had gradually established themselves on land, especially on the eastern coast of our island. It was against the attacks of these restless assailants that the remotest province of the empire had been put in a state of defence at the moment when the Roman legions were preparing to abandon it. The remains of our Roman fortifications from Lympne, in Kent, to Brancaster, in Norfolk, attest the vigilance of Theodosius and Stilicho, and the anxiety of the imperial government to secure the transport of British produce to the Continent. But it was by the admission of these marauders to a share of the soil, rather than by forbidding their approach to it, that the peace of the country could be secured. The coast from Sussex to Norfolk became known in the later age of the empire as the Litus Saxonicum, or Saxon Shore, from the number of this people who had established themselves upon it. We may suppose how in after-times these newcomers aided the invasion of the Angles and Saxons, and contributed to the expulsion of the native Britons from the greater part of our island altogether. It seems that from this period the character of the Saxons as an eminently maritime people underwent a change. In their original habits they were succeeded by the Danes and Normans, people of kindred origin, and marauders such as they had been;

but the Saxons themselves became now a continental and inland people, whether in Britain or in Germany, and adopted the settled political habits belonging to that condition. The Saxons of Lower Germany seem to have maintained a regular polity. They had their three classes, or castes; the Ethelings, or nobles; the Freylings, or free men; and the Lassen, who were their freedmen and clients. Each of these castes kept itself apart from the others with almost Oriental strictness, intermarrying only among themselves, while at the same time they all associated with one another in matters of common interest, assembling every year in a national parliament at a place which is called Marklo, on the banks of the Weser, to take counsel together. The family was firmly based on the institutions, property, and hereditary descent, the true foundations of national power. During a time of peace every citizen of the commonwealth dwelt on his own estate, under the protection of a judge elected by the district. Three chiefs enjoyed a limited authority to convene the men of the three great divisions of the whole Saxon territory above mentioned. When the Saxons engaged in a general and national contest, they chose by lot the commander to whom all should subject themselves in common. In their military array they exhibited a simple but vigorous organization, when their warriors, with their long hair and their short cloaks, armed with a long lance, a small shield, and a knife or axe, flocked around their sacred standard, on which were painted the figures of a lion, a dragon, and a flying eagle.

The same organization which gave strength to the nation secured the vitality of its religious ideas. Their priests contributed another separate caste, and in the midst of an armed nation were forbidden the use of arms, which of itself sufficed to invest their character with a peculiar sanctity. It does not seem clear whether they had learnt to construct special edifices for the celebration of their rites or the use of the worshippers; but they consecrated trees and fountains, and erected altars to their divinities, and on these altars they offered bloody sacrifices even of human victims, which they seem to have sometimes even devoured at their banquets. The central spot of their religious worship was at a place now called Eresburg, near the banks of the Weser, where they set up what seems to have been a rude pillar of stone rather than a statue, to which they gave the name of Irmin Sul, "The pillar of the world." Here they brought the spoils taken in war, and heaped up images of gold and silver before they divided them among their warriors. It was their common practice to sacrifice a tenth of their captives. Their superstitions and their cult, as described in the ancient chronicles, seem to have been generally identical with those reported of the tribes which occupied the same seats in the early days of Germanicus and Varus.

In the course of time the Franks had migrated westward, and had there established an empire founded upon the bases of the later Roman civilization and the later Roman religion. The Saxons, who had succeeded to their inheritance in the North-East, were still, as we have seen, utterly barbarous. A deep, long-settled enmity existed between the two nations, though of the same race, and originally similar in their habits and institutions. They were at constant war, one against the other. But the Saxons, who had begun by driving the Franks across the Weser and the Rhine, had now been driven back in their turn, and with difficulty defended themselves against the encroachments of the Austrasians. Sometimes they were beaten in the field, and required to pay an annual tribute of whole herds of oxen or horses. Again, they revolted against the feebler kings of the Merovingian race, and it was only by the vigour of Charles Martel, and of Pepin, that they were again reduced to submission. They regarded the Christian missionaries as the advanced guard of their political enemy, and saw in the Christian Church established in its bishoprics and its convents on the frontier, the base of an armed invasion, directed against their religion as well as their liberties. When Pepin, and after him Charlemagne, allied himself openly with the Church, making a sort of offensive and defensive compact with it, their religion and their liberties seemed doomed to speedy extinction.

We naturally feel some touch of compassion for, and even of sympathy with, a race from whom such sacrifices are imperiously demanded, even by the law of moral and spiritual progress. We must allow, however, that the so-called liberties of a people like the Saxons were no other than an ill-cemented barbarism, which might well be replaced even by the imperfect culture of the semi-Romanized Franks, while the advance of the Christian faith, with all the defects and corruptions which attached to it under the *régime* of the Frankish Church, was a great and important triumph in the cause of spiritual development. With the wars waged by Charlemagne against the Saxons we need not here concern ourselves further. We have already spoken of them as holy wars, designed ostensibly for the promotion of the Christian Church and faith. As such they did not escape from the fate which has always attended wars of so questionable a character. They were undoubtedly bloody and faithless, even beyond the license so generally assumed by contests which are merely political. They were continued with various success through a period of thirty years, the conquest being completed in the year 804. Charlemagne was too much occupied with other matters of greater, or at least of more pressing interest to his wide-extended empire, to devote the whole of his attention to the conduct of affairs at its remotest frontier, or to concentrate against one enemy the forces required to defend himself, or to advance his projects on all other quarters. From time to time the Saxons proffered submission, and accepted baptism, as the first condition of peace; but again and again they rushed to arms, retracting their enforced vows as a prelude to another struggle. There can be no doubt that the direst acts of cruelty were committed on both sides; but the massacre of the Saxon captives at Verden, to the number, it is said, of 4,500, stands prominent among all the deeds of blood and treachery that have been at any time or anywhere committed; and shocking it is to think that such a deed was perpetrated in connection with a presumed propagation of the Gospel.

The patriotic spirit with which this valiant people resisted the imposition of a spiritual yoke, recommended to them by a foreign and a remote power, such as Rome, in the interest of a foreign power on their own borders, may well command our admiration. We cannot fail to connect it with the spirit of resistance to the same ecclesiastical power at a much later period, which produced the great Reformation of religion under the influence of the Saxon Luther. It was with the greatest difficulty that the free and independent Saxons of the eighth century were reduced to that submission to the Romish tyranny which they finally threw off in the sixteenth. We may admit that the practices they clung to at the earlier epoch had, in themselves, no claim to our sympathy; nevertheless, such as they were, they had been transmitted to them from a remote antiquity; their nation had been bred in them, and they had learnt to bestow some rude reverence upon them. It may be feared that in the new usages which were forced upon them, they found much that was unintelligible, much that appeared merely formal and heartless, much that savoured of human vice and weakness under the specious guise of a divine revelation. It is said, indeed, that there was no ordinance of the Church to which they offered such pertinacious resistance as the payment of tithes, and none, it is believed, was more peremptorily demanded of them. Barbarous and cruel as they were themselves, they were perhaps much less indignant at the barbarity of the penalties that were inflicted by the

Church, and the State which co-operated with it,— death for the burning of churches, death for non-observance of the prescribed fasts, death for the cremation of their dead according to their old national usage, and fines, if not death, for persistence in many other heathen ceremonies. It has been well remarked indeed, that in this, as in other cases of the kind, the effect was not to eradicate the evil practices, but rather to introduce them, more or less disguised, into the very system which so cruelly proscribed them. The Church became itself corrupted by the subtle infusion of the poison which it sought openly to eradicate.

We may remark, however, with satisfaction, that the march of the conqueror's armies, with all the barbarity which accompanied it, was followed more or less closely by the arrival of missionaries of a higher and holier type. We may believe that many of the pious monks who went forth from Fulda, already founded on the frontiers of Saxony, were imbued with the spirit which reigns in the letters of Alcuin, the monk of York, which he did not hesitate to impress upon the imperial patron of the victorious Church. "Faith," he said, "must be accepted voluntarily, and cannot be enforced. A man must be drawn to it, he cannot be compelled to accept it; you may drive men to baptism, but you cannot make them take a single step towards religion. Therefore it is that those who would evangelize the heathen should address them prudently and temperately; for the Lord knows the hearts of His chosen ones, and opens them to understand His word. Even after baptism indulgent precepts must still be delivered to weak souls. So the Apostle Paul writes to the infant Church at Corinth, I have given you milk and not bread. Bread is for men; and represents those great spiritual counsels which are suitable to minds already trained in the law of the Lord; and as milk is for tender age, so we should offer milder rules to the ignorant peoples who are yet in the infancy of the faith. Had the light and easy yoke of the Saviour been announced to these unbending Saxons with the same persistency as the injunction to pay their tithes, and to submit to the penalties imposed upon their slightest faults, perhaps they would not have retained their horror of baptism. Let the preachers of the faith, then, learn by the example of the Apostles; let them be preachers, and not spoilers; and let them trust in Him of whom the prophet bears this witness, that He will never abandon those who hope in Him." Such was the teaching of one, at least, of the most eminent of the ministers of the Church in those days in reference to the conversion of the Saxons; and the same spirit shines forth not less conspicuously in the letters with which the same Alcuin impresses upon the emperor the right means of making due impression upon the Avars, upon whom he was equally resolute in forcing the Gospel at the eastern frontier of his dominions. Again and again his sage monitor urges it upon him to let the heathens be taught the Gospel before they are required to accept baptism; again and again does he warn him that belief in the saving truths of religion is of more importance than the payment of tithes;

it may be further remarked that in Alcuin's summary of the duties of a Christian there is no mention of his submission to any specific ecclesiastical system. Nor are there wanting some faint traces, at least, of similar appreciation of the moral teaching of the Gospel in the counsels which these missionaries received from Rome itself, and we may hope that the bloody instincts of Charlemagne and his Frankish warriors were tempered in some degree by the gentler and wiser spirit of the monks who followed in their train. It may be feared, however, that even among the genuine teachers of the true faith there was too much disposition to disregard the sufferings of their actual generation in view of the triumph that would ensue to the Church in the subjection of their children and their grandchildren. The persistence, indeed, with which the conqueror carried out his policy began to bear fruit in his own time. As the resistance of the daunted barbarians gradually relaxed, he was enabled, on his part, to relax from the unrelenting severity with which he had continued so long to enforce it. It is observed, that from the date of the year 797 the laws or Capitularies which he promulgated in reference to the Saxons assumed a milder form. The principles of the native law—for the Saxons, barbarians though we commonly designate them, had a rude code of laws of their own, even at this early period—were respected and confirmed, and their nationality thereby assured to them. Six years later a compact was made between the conqueror and the conquered at a meeting held at Salz. On the one hand Charlemagne, who had now assumed the title of Roman Emperor, allowing the pope to place the crown upon his head, and who had invested himself thereby with all the grandeur which still attached in universal estimation to the inheritance of the Cæsars, the masters of the ancient world, on the other, the chiefs or nobles of the Saxon confederacy, brave and spirited men with arms in their hands, might stipulate with one another on some footing of equality. The Saxons promised to renounce the worship of idols, to submit to the instructions of the bishops in matters of faith and moral practice, and to pay with due punctuality the tithes prescribed, as they were assured, by the God of the Christians. The emperor, in his turn, engaged to relieve them from every kind of tribute to the imperial treasury, to preserve their liberties, and to leave them the government of their own affairs, reserving only to himself the right of appointing their judges, the acknowledged representatives among the Germans of the royal person and prerogative. He took the precaution, however, of requiring hostages for their faithful observance of the conditions he imposed upon them.

The public act by which the religious organization of the new province of Christendom was effected dates, indeed, from a diet held at Spires in 788, when the subjugation of the Saxons was as yet prematurely anticipated. Its terms, however, may be cited as apparently identical with those which were finally ratified fifteen years later. It "notified to all the faithful in Christ that the Saxons having, notwithstanding their obstinacy and perfidy, been, at last, by divine favour, subdued and brought to baptism, Charles the king had restored to them their ancient liberty, and for the love of Him who had given him the victory, had surrendered them to Christ as His tributaries and subjects. Wherefore, on reducing their territory to the form of a province, according to the ancient use of Rome, he had divided it among several bishops, the first of whom should be established in a place called Bremen." Seven other sees were erected at the same time, at Osnaburg, Paderborn, Munster, Minden, Verden, Hildesheim, and Halberstadt. Each see presented to God an altar, to Truth a pulpit, to Justice a tribunal, to Charity an asylum, to every idea of benevolence an institution which might enable it to penetrate into the manners of the people. Around these episcopal seats parochial churches gradually grouped themselves, and conveyed the same ideas, maintained by similar institutions, to every corner of a country which for so many ages had been given up to ignorance and the law of the strongest. The eloquent and feeling writer from whom these details are borrowed, himself a devoted servant of Rome, goes on to remark upon them, that thus "the war in Saxony, *for a moment compromised by the error of the temporal power*," that is, by the worldly ambition of Charlemagne, "seemed to justify itself by its results, and, as has been the case with all holy wars, has served the cause of civilization. Nevertheless, the conqueror's conscience would not have enjoyed repose had he been allowed to witness the result of his work, which was destined to manifest itself seven centuries later, at the breaking out of the

Reformation. The Roman faith, while remaining mistress of the populations of Frankish and Bavarian origin, among which it had been established simply by the power of preaching and of Christian charity, was then betrayed by the descendants of the Saxon tribes whom the soldiers of Charlemagne supposed themselves to have conquered. And who shall say whether Luther himself, the son of the miner of Eisleben, did not spring from the blood of some one of the four thousand five hundred captives whom they had massacred at Verden?"[1] It may be hoped that the touch of remorse here remotely indicated may be something more than a mere rhetorical effusion.

We may be permitted, however, to point out as a surer foundation for the independent spirit of the Saxon Christians, that it was from the preaching of the Anglo-Saxon or English monks, for the most part, that they had received their earliest instruction in the vital principles of the faith. They inherited, we may believe, through many generations, the spiritual courage which continued from age to age to distinguish the professors of our faith in England from the Christians of the continent generally, and especially those of southern Europe. Rome could boast, indeed, no more devoted children than the English as long as she stood before them as a pure source of spiritual enlightenment, as a propagator of genuine Christian principles, and of the civilization which never fails to follow them. But when she manifested herself to

---

[1] Ozanam, 'Études Germaniques,' ii. 262.

the world she had subjugated as a mere worldly power, pretending to a divine mission, which she had forfeited from her proved unworthiness, from that moment all that was truly Christian in the English Church rebelled more or less openly against her, and asserted the indefeasible right of the believer in Christ to follow Him and Him only, casting off all fallible human guides, and trusting to His loving guidance to lead into all necessary truth. It would be more reasonable, perhaps, to imagine that the son of the miner of Eisleben was induced to revolt from the tyranny of Rome, not from any instinctive sense of the injuries done by Charlemagne to his forefathers, but rather from the spirit with which he had been himself imbued by the long-descended traditions of the Anglo-Saxon discipline. The missions of the Anglo-Saxons had replaced in the Carlovingian epoch the Irish missions of the Merovingian. They had been directed equally to the conversion of the heathen and to the reform of the professing disciples of the North. Commencing from the time of Pepin Heristal, the father of Charles Martel, early in the seventh century, they had been carried on effectively by Winfrid, Suitbord, and Willibrord, in Frisia; by two monks named Ewald in Saxony; by Boniface, who, as we have seen, after his long wanderings had fallen a martyr on the shore of the German Ocean. Alcuin, of York, whose godly spirit has already been noticed, exerted his influence upon those northern missions from the centre of France, in which he had planted himself. The purity and simplicity of the English school of teachers contrasted favourably with the worldly character of the Frankish priesthood, and Charlemagne himself was impressed with the importance of intrusting the establishment of the Church throughout his northern conquests to these foreigners rather than to his own subjects. He appointed the Anglo-Saxon Willibrord to preside over the district of Estphalia, and Liudger, a Frisian by birth, but an Englishman by his training at York, to organize the Church in Westphalia; while he left to the earlier foundation of Fulda, which had also received its first Christian traditions from the English Boniface and his pupil Sturm, the charge of Engern or Angaria. From the teaching of these strangers there sprang up a crop of Saxon priests and missionaries; from among the youths of noble family whom the conqueror had carried off from their homes as hostages, many were selected to be trained in the monasteries for the life of monks and preachers. Eventually the Abbey of Corbie, near Amiens, was founded by one of these Saxon converts, and became an important centre of Christian teaching. From hence sprang the daughter-foundation of the New Corbie, or Corby, on the banks of the Weser, in the diocese of Paderborn. This abbey received its charter from Louis le Debonnaire in 823, and became no less important an institution for the propagation of the faith in the north of Germany, than Fulda still continued to be in the centre, and St. Gall in the south.

Meanwhile the Faith continued to advance quietly along the southern shores of the Baltic, chiefly through the missionary zeal of preachers from central Germany. Among these, Bishop Otho of Bamberg is signalized as the apostle of Pomerania, in the twelfth century. But the final conversion of North-Eastern Germany was really effected at a still later period by the knights of the Teutonic order, the origin of which cannot be traced beyond the year 1190. During the siege of Acca, or Ptolemais, by the Crusaders, certain pious traders from Bremen and other German sea-ports devoted themselves to the service of the sick and wounded. The princes of their nation rewarded their zeal by enrolling them in a military fraternity, and they were sworn to the defence of the Faith, as well as to the relief of the poor in the Holy Land. When the Christian forces were expelled from thence, these Teutonic knights retired home; but they now occupied themselves with the conquest of Prussia, Livonia, and Courland, and propagated their religion with the sword. This crusade lasted for one or more centuries, but it ended in the complete subjection of this large tract of country to the knights, who became its secular rulers, with the sanction of the pope on the one hand, and of the German emperor on the other.

CHAPTER IX.

MORAL INFLUENCE OF THE SECULAR EMPIRE IN THE CONVERSION OF THE NORTHERN NATIONS.

THE Teutonic nations which infested the northern frontier of the Roman empire could not fail to acknowledge a traditional apprehension, derived through many generations, of the mighty military power against which their furious assaults had been so repeatedly broken. From age to age they persisted in warring one against another; they were chafing within the bounds, all too narrow for their roaming disposition, of their own respective territories; sometimes gaining, sometimes losing ground among themselves, they might observe a general tendency throughout their races to press blindly onward to the south or west, till they felt that there was an obstacle before them, a chain of mountains, a mighty stream, a line of fortifications, at which their progress was arrested by an insuperable resistance. From time to time, indeed, this limit was partially overcome or evaded; now and then a sudden victory gave admission to a portion of the great barbarian horde within the sacred boundary, and a section of one of their tribes became permanently planted on Roman soil beyond it. But such cases were rare and of little apparent moment; they made themselves only partially known beyond the spot where they actually occurred. The great bulk of the Northern barbarians was still wont to regard these Roman defences as impregnable. Nor were there ready means of discovering what it was that lay behind them; what were the actual forces, and how great the ultimate resources of the political body which they thus protected from approach. The communication between the Romans and the Teutons was too slight, even on their borders, to allow the barbarians much insight into the secret of the Empire, the secret of the strength she ordinarily opposed to them, still less of the strength which on any serious pressure she could rally for her protection. The power of the great Southern people was a mystery on which they ruminated, perhaps, in their own recesses, which awed their imaginations, and filled them, no doubt, with an idea of its moral superiority, subduing their courage and abating their presumptuous ambition.

But the power of the Empire continued actually to dwindle, and when more and more detachments from their armies began to be admitted within its borders, and bribed to defend it, the Northern races perceived at last its intrinsic weakness. Yet they were not insensible to the advantages it still possessed in the completeness of its social organization, in its civil and military administration, in the genius for government by which it made all its forces work in unison and bound together all the members of its vast community. When, in the fourth century, the Teutons had acquainted themselves with the play of the great machine before them, the Empire had become almost wholly Christian; it was clear, at least, that all its effective strength resided in the large and increasing element of its Christian population, wherein all its vital principles were centred. Amidst the internal differences which agitated its Churches there still reigned an external unity and harmony, which doubtless made a deep impression upon the rude, independent, and mutually repellent elements of barbarian society. The barbarians could never act in concert; the Christians were always associating and combining together. Nothing, for instance, could have raised more curiosity and wonder among the Goths and Germans in the provinces than the spectacle of troops of Christian prelates flocking towards some central spot along the roads of the world-wide Empire, to deliberate concerning points of common interest, points of difference or agreement, upon which their combined action should afterwards depend. We must never forget that the great Œcumenic councils, such as those of Nicæa and Constantinople, presented a spectacle of which the world had had no previous example; a spectacle which would have astonished the ancient Greeks, civilized as they were, and accustomed to meet in their national assemblies near to their own doors; which would much more have astonished the Romans of the highly-organized society of Augustus or Trajan, who had never been able to conceive the working of a representative body. To the Northern races, whether within or without the borders of the Empire, the meeting of these Christian deputies for common counsel together, from points some thousands of miles apart, was the most startling of phenomena. It could not fail to impress them with profound awe and admiration of the institution which could wield so quietly the power, and so surely mould the will of the whole mass of civilized life among men. We shall not err, perhaps, if we attribute a higher influence on the popular imagination to this evidence of spiritual power than to the sight even of the material splendours of the Empire, its cities and palaces, its roads and bridges, or even its long array of armies and the memorials of their innumerable triumphs. If Rome had lost, in the fourth century, some of her grandeur as a conquering state, she had acquired more than a compensation in the eyes of the new immigrants in her character as an agent of human culture. The Northern peoples were not unacquainted with arms, but they were as yet wholly strangers to moral and material refinements, and these were the forces which were now brought most prominently before them.

Nor was the interest of these strangers abated when they heard of the trials and sufferings through which this marvellous spiritual organization had worked its way to the prominence and efficiency which now so manifestly attached to it. The Council of Nicæa was attended by at least one bishop of the Goths, and he could bring back to his countrymen on the banks of the Danube the story of this meeting of chosen delegates from three hundred provinces, most of whom had witnessed, some had suffered under, the torments of persecution but a few years previously. He could tell them how he had seen and spoken with Paphnutius, a prelate from the Thebaid, when that venerable professor of Christ had entered the hall of council, trailing under him a limb that had been hamstrung, and presenting the hollow orbit of an eye that had been torn out; how he had received the blessing of Paul, bishop of Neo-Cæsarea, on the Euphrates, given with a hand which had been maimed with fire; how he had joined the multitude there assembled in prostrating himself before those noble witnesses to the truth, and in kissing their scars and wounds. Nor was it such sufferings only that these holy men attested: there were among them many who had come from the monasteries in the deserts of Egypt and Arabia, who had passed their lives in the practice of asceticism in solitude, who had proved by their own example that the Christian requires for his moral and spiritual sustenance no communion but that with Christ and the spirits of His saints in heaven. Nor, again, were there wanting others who represented the happy effects of godly living, even in the crowded cities of the world around them, men who were noted for being in the world, but not of it, noted for their austerity, their purity, and their learning. The stranger from the confines of heathen barbarism was constrained to acknowledge that the best and wisest of the generation into which he was himself born were men devoted to the Christian persuasion, men who showed forth in their lives the power which the Gospel was now manifestly exerting upon the spirits of mankind, the power which it might haply be expected to exert more and more upon the generations that were to follow. The grave questions of theology which these representatives of the Church of Christ were met to solve might seem to touch upon the

MORAL EFFECT OF THE COUNCIL. 139

deepest interests of humanity. They referred to nothing less mysterious than the very nature and essence of the Deity. The question of questions, which had been held in suspense for so many ages by every sect of religious or philosophical inquirers, seemed about to receive authoritative solution under the guarantee of a divine revelation. God Himself was about to discover Himself by the mouth of infallible interpreters. It was an awful moment to the devout believer. The council of the bishops at Nicæa might well have kindled the imagination of the bystanders, even of such as were invited to watch and hearken to it from the confines of darkness and unbelief.

The effect thus made upon the anticipations of the new races which were hovering on the frontiers was well calculated to balance the apprehension they might entertain of the impending fall of the ancient empire. Throughout the course of the fourth century the organization of the Roman government continued to show signs of approaching dissolution, disguised, indeed, by occasional revivals and moments of feverish vigour, but generally none the less patent to the ardent peoples who were impatient to seize upon its inheritance. The pagans of the Empire still persisted in proclaiming the eternity of Rome, of her polity and her moral authority over the world. They asserted defiantly that the existence of the fabric of human society depended upon the maintenance of Rome as the head of the whole social system. The world, they contended, was made for Rome. This persuasion, impressed deeply upon the pagans, they had contrived, by the dogmatic assurance with which they held and enforced it, to impress even upon the Christians also. The Church herself held that from the capital of the Empire went forth the spirit of human culture, both material and moral. To no other city, to no other metropolitan centre, did she attach such pre-eminence among the divine instruments for the education of society; not to Athens, or Alexandria, or Antioch; not to Constantinople, however fondly its founder might hope that she would supply the place of Rome, at least to one half of his empire. Rome, even in the eyes of the Christian believers, even in the fourth century, was still the divinely-appointed source of human civilization.

The gallant career of the Emperor Theodosius arrested for a space the downfall of the doomed city. The interval was providentially designed for the secure establishment of the Christian Church as the accepted religious teacher of the Empire, before the ancient stronghold of paganism should, in the fulness of time, be dislodged from its imperial pre-eminence. Rome was at last taken by the Goths under Alaric. Three times did these barbarians from the North appear before her walls. Twice they retired after extorting a ransom, and after placing a puppet of their own choice on the throne, in opposition to the emperor, who had taken refuge at Ravenna. The third time, in the year 410, they entered. The gates had been opened to them by the treachery, or rather by the terror, of the unwarlike population, and though the city was subjected for six days to the greed and violence of the conquerors, there was no general spoiling or massacre of the inhabitants. But the impression which this awful disaster made upon the pagans, and the Christians also, was not on this account the less signal. The pagans regarded it as the overthrow of their faith, together with their temporal dominion. To their imaginations it sounded the knell of the old world, to which they could anticipate no successor. All beyond it was chaos. The Christians seem themselves at first to have had in these overwhelming apprehensions. They supposed that the very existence of their own spiritual Church was interwoven with the framework of the civil society with which they had been so long and so closely connected. Their apologists in an earlier generation had accepted the dominion of the conquering city as a principle of the Divine government of the world. The rule of the emperor was to them a law of Providence. So they interpreted the Lord's injunction, to render unto Cæsar the things that are Cæsar's, and to God the things that are God's. The world in their view was pagan or secular, as contrasted with the society of the Christians, or the Church of Christ, which enjoyed the promises of the future, but had no share in the enjoyment of this present life. They could not fail, indeed, even in the flourishing age of Constantine or Theodosius, to see that Rome was environed with perils, and to apprehend that her days were numbered; but beyond the fall of Rome they could imagine no future, unless it were the appointed end of the world, which they had been so long expecting. As years went on and these perils thickened, this was the idea which impressed itself most strongly upon the minds of the Christians. Here, at least, the adherents of the two religions were in accord. Both believed that Rome was necessary for the world, and would last as long as the world lasted, and perish with it.

The barbarians, on their part, had been long wont to regard the imperial city as the centre of a mighty power, which they could neither measure nor appreciate. It had long inspired them with awe, not for its military prowess only, but for its moral grandeur. It had impressed upon them a sense of law and polity, of the abiding power of mind over matter, of moral over material forces. To the mass of the Northern barbarians on the frontier, Rome was still the embodiment of government and order. But of the Goths who had been settled within the Empire, many had by this time become imbued with the rudiments of the Christian faith. A certain number of them at least had been admitted into the Church by baptism; they had learnt to invoke the Holy Name, and receive it, as far as they considered such dogmas at all, under the definitions prescribed by the Court and Church of Constantinople. We may well believe, accordingly, that by the barbarians from the Danube, Constantinople was already regarded as the centre of the Empire much more than her elder rival. In their view, the fall of the ancient capital was an event so much the less striking, as they had never learnt to look upon it with the same eyes as the races of the West. To their apprehension, the Roman empire was in no such danger of immediately collapsing with the destruction of Rome, much less with its temporary subjection to the stranger. They could still turn to the new metropolis in the East as the centre of the imperial power which they had so long respected, and regard it as the true source of law and order, of moral and intellectual culture. To Constantinople they were now encouraged to look as the spiritual mistress of the nations. Providence had not withdrawn its protection and favour from the sphere whence the rays of revelation had first penetrated to their own dark abodes, whence they were doubtless destined to travel onwards throughout the heathen world. Perhaps their own Mother Church would shine in their eyes all the more resplendently from the extinction of the rival luminary in the opposite hemisphere.

But the fall of Rome, as was soon discovered on all sides, was not the crowning catastrophe which had been so widely anticipated of it. The Northern invasion swept through its streets and rushed forward in pursuit of other victims. The population, scattered by the shock, returned, though in diminished numbers. The buildings were restored, though with diminished splendour. The apologists for the faith who followed in the light of the event itself, declared that the fall of the secular head of the world was the rise of a spiritual head, which had been from the first appointed to succeed it. If not in this world, assuredly in the next, the City of God was destined to take the place of the City of Man ; the overthrow of the Empire in the West, which had been its cradle and its throne, was, after all, necessary for the full development of the spiritual dominion of Christ, of which the prophets and the apostles had spoken.

St. Augustine, in his famous treatise on the 'City of God,' stepped forward to soothe the excitement of the believers, by first pointing out that the calamity in which they had shared with their unbelieving brethren might be regarded as a chastisement for their own defects, and a warning of God's power and justice. From thence he expatiated on the proof it furnished them of the vanity of the pagan notion that Rome, the pagan city, was protected by a special Providence. Often had she been afflicted aforetime by war, pestilence and famine; now, at last, she suffered a crowning judgment on her wickedness and unbelief. And then he invited all mankind to accept the revealed will of God in Scripture, and look to Him as the protector, not specially of empires and worldly governments, but of all devout believers in Him and His Christ.

This spiritual interpretation of a spiritual dispensation exerted a marked effect upon myriads both within and without the limits of the Empire. We cannot doubt that it opened to the German races within, and even beyond, the Danube, a new view of the intrinsic force and purport of the Gospel. It confirmed multitudes in the faith they had already partially and imperfectly apprehended, and it purged the vision of some who had hitherto regarded it merely as an adjunct to the secular empire, an appendage to the skirts of the court at Constantinople. The time, indeed, was not yet come for the full conception of the grandeur of Christianity itself, or of the demands its providential career would eventually make upon the acceptance of the heathen.

But Constantinople had meanwhile some great advantages over Rome as an educator of the human race. As such, she was in fact the successor of Greece rather than of Italy, of the Greek character rather than the Latin. The intelligence she now called into play was essentially Hellenic, and largely partook of the interests to which that intelligence had been in the course of its earlier career most strongly directed. Thus it was that throughout the great cities of the Eastern Empire no intellectual speculations were for many generations so rife as those on the nature of the Godhead, such as had engrossed the attention of the ancient Greek philosophers from the dawn of thought, but which were now confined within the strict limits of a volume of sacred texts. The learning and the subtlety with which these written texts could be handled afforded an unfailing source of interest to the divines and preachers who undertook to interpret them; while at the same time they inspired their hearers with admiration, and often with the blindest devotion to their teachers. Signal, no doubt, was the effect upon the ruder intellects of the Northern nations, still undergoing their novitiate in the Church of Christ. The rapidity, for instance, with which conversion followed upon the teaching of an Ulphila, shows how ready the Goths and Germans were to yield to the demands advanced upon their faith and reason. These strangers looked more especially to Constantinople as the source from which their new learning flowed, from which the doctrines issued which were declared to them as necessary to their salvation. Accordingly, we may readily imagine the hold which the city of the Cæsar, and the person of the Cæsar himself, would obtain upon their imaginations. They promptly accepted, with little or no examination, the phase of Christian belief which was so speciously presented to them. The general entertainment of the Arian tenets by the converts from the North at a time when among the more advanced sections of Christian believers there prevailed a great diversity of opinion upon them, and when, indeed, the orthodox party, though less in favour at court, still numbered an undoubted majority throughout Christendom, is one of the most curious circumstances in the history of the Christian Church.

There can be no doubt that, at this period, the influence of Southern civilization upon the imagination of the Teutonic races was powerfully assisted by the literary resources which it continued to unfold. If the hands of the great conquerors were now weaker than of yore, and most plainly so in comparison with the fierce barbarians who were disputing the Empire with them, the mind and intellect they displayed on all points of human interest would suffice to secure for them a moral superiority. To no point was the mind of the age more strongly directed than to the study of theology. The Gospel was preached as a theological science, rather than as a rule of practical discipline; as a passport to a future life, rather than a preparation for it. The Christian controversies of the day were mainly engaged, as we have seen, on the problem of the Divine nature, a problem which of all others was the fittest to subdue the intellect of the pagan inquirer, and induce him to submit to the teaching of doctors who claimed to be wiser and better instructed than himself. He heard from afar off the echoes of the great preachers at Constantinople, and other centres of Christian science, who expatiated so constantly on the nature of God and the distinction of persons in the Deity; he saw with his own eyes the train of dignified ecclesiastics who rushed from all quarters to meet in council upon these and similar dogmas, while priests and people waited in breathless suspense upon their decisions; he received, from time to time, some missionary teacher who came to him with a declaration of the faith, to which his unquestioning adhesion was imperatively demanded. Thus plied on all sides, and taught to believe that on his reception of such tenets his salvation depended, he still felt that the Empire of the Romans was the appointed mistress of the world, and bowed to her spiritual authority even more humbly than of old to her military pre-eminence. We may remember also that the ecclesiastical literature of the fourth and following centuries was probably not less Arian than Athanasian. We have lost the vast amount of writings and preachings which enforced the heterodox tenets, while the orthodox replies have alone been suffered to remain to us. The Northern nations, on their admission within the pale of the Church, were addressed at least equally by both parties; and of these we are given to understand that the heterodox disputants were the more popular, and made themselves more generally intelligible than their opponents. They worked, indeed, on a lower level, suitable, it may be presumed, to a lower order of minds. The barbarians, we may believe, regarded them as the more powerful faction of the two; but it was rather, perhaps, as exponents of the power of the Empire than for any special powers of persuasion, that they so generally preferred their teaching.

But whatever might be the force exerted by these moral influences upon the minds of the heathens from the North, we shall be inclined to attribute still greater effect to the support which the State so ostentatiously lent to the Church in the Eastern Empire. The union of the Church with the State, which was first foreshadowed in the policy of Constantine, had become in the next generation complete. The rulers of the Church were, for the most part, more or less directly appointed by the head of the secular government; he it was who called them together to take order on questions of discipline and doctrine; sometimes himself in person, at other times by a deputy of his own nomination, he presided at their deliberations. The appearance of Constantine surrounded by his guard at the Council of Nicæa has been specifically described to us, as from its historical importance it deserved to be. The emperor promulgated the ecclesiastical decrees, and enforced obedience to them by the sword of civil justice. The legislation of the Empire was moulded upon Christian principles, and bore many significant marks of the new ideas introduced by them. The emperor was recognised, in fact, as the head of the Church upon earth, and the half-divine character with which he had been invested by the superstitious adulation of his pagan subjects, still continued to float, as it were, over him, and entice his willing flatterers to ascribe to him somewhat more than merely human attributes. When it was asserted of Constantine, as appears to this day in the inscription on his Arch of Triumph, that he subdued his adversary through the operation of a Divine Influence (instinctu divinitatis), both pagans and Christians equally recognised in this bold expression a reference to the claim of the secular ruler of the Empire to be animated by the Spirit of God.

Doubtless the intelligent Christian and the polished pagan might equally extenuate the meaning of this rhetorical language; but to the simpler child of the Northern deserts, to whom it was interpreted at second hand by superstitious or crafty priests, it represented a plain truth to which he must bow in faith and reverence. When the chief of the State was thus formally set before him as an inspired instrument of God, even as closely allied to the Divine nature himself, he could not but acknowledge that the faith which such a chief dictated to him had received the highest sanction of which it was capable. The Gospel was thus recommended to him by a power which might lawfully demand his acceptance of it. He was not only disposed to the faith, and subdued by its arguments, he felt a moral compulsion placed upon him which few could be so bold and reckless as to disregard. As the people of Judah and of Israel, in the olden time, had readily accepted the worship imposed upon them by their sovereigns, so the races which hovered on the frontiers of the Empire, or obtained admission within it, accepted Christianity at the beck of the emperor. They accepted not the general principles of the Christian faith only, but also the distinctive speculative tenets which were from time to time recommended to them by his precept or example. What was the real nature of such a faith, what the depth to which its principles penetrated, it might be painful to inquire; but the main fact is beyond dispute, that the conversion of the Northern nations on the borders of the Eastern Empire was very powerfully influenced by the visible alliance of the Church with the State.

CHAPTER X.

INFLUENCE OF THE ECCLESIASTICAL SYSTEM ON THE CONVERSION OF THE NORTHERN NATIONS.

THE moral influence of the civil government of the Empire was confined, for the most part, from the fourth century onwards, to the eastern portion of Christendom. We have just seen how important a part was played by the emperor at Constantinople, and the worldly influences of his court around him, in the propagation of Christianity among the races which descended from the North. We may now turn to the West, and notice the very different conditions of political and social life under which the same divine work of conversion was carried on by Providence in that quarter, till it issued in the general establishment of the Church of Christ throughout the civilized portion of Europe. Hitherto the external causes which have been principally at work have been connected with the employment of the secular power; we shall now find the place of the secular taken by the ecclesiastical system, the Bishop of Rome, or the Pope, as he may be from thenceforth called, will exercise an authority over the Church, not only spiritual but political. For the union of the Church and the State will be now complete; the two powers will have become conterminous; the Church, truly Catholic, will embrace the secular together with the spiritual element, and the policy of both will run for the most part in the same channel.

When Constantine had withdrawn to his eastern capital, the balance between the old and the new Rome was not long preserved. The Roman Senate still occupied its ancient quarters; while it claimed still to exercise certain unsubstantial prerogatives, it received, perhaps, some addition to its power or lustre from the absence of its imperial masters. But Rome began soon to show signs that she had lapsed into the position of a mere provincial city. When the severance between the East and West came to be effected, Honorius planted his throne at Ravenna, and it was from a corner of Italy rather than from Rome that his decrees were dated, and the civil authority emanated. The traditions of the old world were felt to be too strong for the control of a government which rested upon a basis of new ideas. Even the political movement of the Empire would have been hampered by the jealousy of the old patrician families, which still exercised a moral supremacy at the seat of their historic greatness; and when the court had become Christian, the still lingering prevalence of pagan associations would have made the emperor feel himself ill at ease in such a locality. The Romans now quickly recovered the lessons of self-government which they had forgotten under the Cæsars on the Palatine. For a time they cherished their pagan predilections, and continued to disturb the ecclesiastical policy of their absent rulers by the exercise of their restored freedom. They had stiffly opposed the measures of Theodosius and Gratian for the suppression of pagan worship. They had remonstrated, and for a time effectually, against the removal of the statue of Victory from their ancient Senate-house. They now persisted, in spite of direct prohibition, in raising altars and shrines and performing their customary worship, and succeeded in postponing for more than one generation the complete overthrow of their proscribed superstition. But in the meanwhile this assertion of freedom was operating not less to the advancement of the ecclesiastical system in the very focus of still lingering paganism. The disciples of the Christian faith learnt the lesson of self-reliance, and freely thought, spoke, and acted for the promotion of their religious views. To these they sought to give effect by putting themselves under the immediate guidance of their spiritual leaders. They claimed as an indefeasible right the election of their spiritual chief or bishop; and this prerogative, which the emperor still frequently exercised *de facto* in the East, was conceded to the people or tacitly abandoned in the West. It was to their bishop rather than to the prefect whom the emperor sent them that the Christian population of Rome betook themselves on every emergency. When Alaric first threatened the city, it was of the Pope, Innocent, that they asked permission, Christians and pagans together, to consult the Etruscan soothsayers. If this strange story, though recorded by writers of both persuasions, be inexact, as it most probably is, there is no doubt that when Attila was hovering on the banks of the Po, it was Leo, the Pope, whom the whole population combined in sending to intercede and make terms for their preservation. No consul, dictator, or emperor, with all his legions at his back, could have done more for the dearest interest of his country than was effected by the peaceful and respectful interference of the spiritual head of the Church at Rome.

"Leo the Great," says Dean Milman, in an eloquent passage of his history, which the reader will be glad to read, even if it be for a second time, "was a Roman in sentiment as in birth. All that survived of Rome, of her unbounded ambition, her inflexible perseverance, her dignity in defeat, her belief in her own eternity, and in her indefeasible title to universal dominion, her respect for traditionary and written law, and for unchangeable custom, might seem concentred in him alone. The union of the Churchman and of the Roman is singularly displayed in his sermon on the day of St. Peter and St. Paul; their conjoint authority was that double title to obedience on which he built his claim to power, but chiefly as successor to St. Peter, for whom and for his ecclesiastical heirs he asserted a proto-apostolic dignity. From Peter, and through Peter, all the other Apostles derived their power. No less did he assert the predestined perpetuity of Rome, who had only obtained her temporal autocracy to prepare the way, and as a guarantee for her greater spiritual supremacy. St. Peter and St. Paul were the Romulus and Remus of Christian Rome. Pagan Rome had been the head of the heathen world;. the empire of her divine religion was to transcend that of her worldly dominion. . . . . It was because Rome was the capital of the world that the Chief of the Apostles was chosen to be her teacher, in order that from the head of the world the light of the truth might be revealed over all the earth."[1] If such, however, were the sentiments of the bishop, they were a reflection of the sentiments which were already dimly brooding in the minds of his fellow-believers. The mission which they devolved upon him was the natural result of the special eminence they attached to his office, and the dignity, still more the success, with which he fulfilled it might sanction in all eyes the principle involved in it. Alaric the Goth had imbibed some tincture of, at least, a nominal Christianity. But Attila the Hun only knew its name as the creed of races which he despised, and of which he proclaimed himself the scourge. But even he deemed himself the scourge of God. He had some sense of a Divine Power moving in his bosom. The sight of the magnanimous prelate who dared to approach his camp with no other attendant but his crossbearer, may have struck him with amazement and admiration; but when Leo reminded him of the fate of Alaric, and predicted the quick and sudden death of the assailant who should desecrate the shrines of the Apostles, he seems to have actually quailed before him. The pretended vision of St. Peter and St. Paul, with threats of this divine judgment, may have been the invention of a later day.

Attila withdrew from the conquest of Italy, which lay so helpless at his feet, and the invasion of the terrible heathens was baffled. The triumph of the

[1] Milman, 'Hist. of Latin Christianity,' book ii. ch. 4.

Church was signal. Providence ordained that in the collapse of the temporal power she alone should stand forth as the mistress and guardian of human civilization. If the heart of the Huns was still hardened against her faith, the Teutonic nations—such, at least, as had come within reach of her preaching—were struck more and more deeply with the spiritual authority she seemed now rightfully to claim. Rome, the mighty conqueror, abandoned by her Cæsars and her legions, had been left to the counsel and protection of her bishops and priests. The clergy might go forth, trusting in no arm of flesh, but in the power of the Holy Spirit, to save the lives of men, the churches consecrated to God, the bones and relics of the saints. They, too, like the barbarians themselves, had been the enemies of Cæsar; they had now become his best allies and defenders, while they invited them to join with them and learn from their lips the gospel of a new life. They taught him the ways of God to the nations, justified his career as the instrument of Providence, and assigned to him his place in the roll of divine revelation. We shall not be wrong in pointing to this great event, the conference of Leo with Attila, as the crisis in the conversion of the Teutonic races, which attached the great bulk of them from this time permanently to the Church, and especially to the Latin branch of the Church, which now reigned in supreme majesty in Rome itself.

The spiritual authority of the Pope had been fully established when the Northern races effected their final conquest of Italy. Theodoric and the Ostrogoths were already converts to the faith when they settled themselves in these new acquisitions. The Arian principles which they had imbibed in the course of their journey westward gave way gradually before the unbending orthodoxy of the Church they found so firmly planted at the centre of Christian illumination. The accession of a temporal ruler, so obsequious to the superior claims of the ecclesiastic, was rather a gain than a loss to the consideration and real power of the Pope. The Ostrogoths became, in a few generations, devoted to the tenets of the orthodox Latin Church, and the spiritual influence of the Holy See, as the Roman bishopric was now popularly designated, was enhanced by the triumph it had achieved, a triumph which extended far beyond the limits of Italy, and dominated over the conversion of many swarms of Northern heathens.

As the emperor had been assisted by an official hierarchy of prefects in every province, so the Pope of Rome, as soon as his superior authority was fully recognised by the bishops and clergy around him, found his nearest and most effective support in the vast array of prelates, among whom all the provinces and districts of the old Roman dominion were now distributed. The number of these spiritual officers at any one period is not accurately ascertained, but, speaking broadly, it has been computed that there were one thousand such in the Eastern, and eight hundred in the Western divisions of the Empire. Wherever the Germans penetrated within the frontier, they were encountered by some representatives of this peaceful army, men of peculiar and dignified bearing, clothed in long vestments of a grave character and colour, carrying, perhaps, a staff in one hand and a book in the other, haranguing the people around them in assemblies brought together at their persuasion, offering advice in kindly tones, and when they had gained the ear of their spectators, unfolding to them matters of deep import concerning their duties here and their prospects hereafter. The creed they taught was a new revelation; but the men themselves were a new revelation also. None like them had been seen by the barbarians of the North; no such missionary teachers had been produced by the cultivated heathens of Greece and Rome before them. Nor did these men preach and teach only; they suffered for the faith they had proclaimed so loudly. Not only did they endure persecution, and subject themselves to martyrdom for the sake of their own souls, or for an example of holy living and dying, to their disciples, but they declared themselves the representatives of a system widely diffused, deeply rooted, centred in Rome, for the sake of which they thus placed themselves in the front of holy battle.

This system they further described as the divine source of wisdom and virtue, as the appointed means both of civilizing mankind and of saving it from its sins. The Germans might behold in these august strangers the bishops of the missionary age; for they came to them almost always with the title and prerogatives of the episcopal order. They avowed that they had been specially appointed to their office as missionaries, in most cases by the Pope himself, to whom they had betaken themselves in order to receive his sanction to their undertaking, to warm the flame of their devotion at the shrine of the Apostles, to obtain the investiture of the sacred robe or pallium from the hands of St. Peter's spiritual descendant. With such credentials to show, they made, as we may readily suppose, a deep impression upon their rude disciples, filling them with awe of the Holy See, and binding them effectually to its yoke. We have traced the progress these missionaries made among the heathen on the borders of civilization, and seen how much of their success was deservedly due to their personal character for faith and devotion; but we must not fail to recognise the mighty power which lay in ambush behind them, the vision of an established theory and system which loomed in the distance, and added a mysterious charm to the words they uttered. The missionary bishops of the North constituted the point of the lance, but the Church, represented most plainly by the Holy See at Rome, was the mighty stem which gave it force and momentum.

The influence of these holy men, thus supported, was no doubt powerfully aided by the contrast their lives and manners presented to those of the only rival class which demanded the submission of the multitude. The Northern races, destitute as they were of political institutions for the security of popular independence, acknowledged before the arrival of these new-comers no superiors but their lords or barons, who themselves owned no superior but their feudal sovereign. These chiefs spent their time mostly in warfare, and it was, perhaps, only this circumstance that saved the missionaries from extermination; for the jealousy of the leaders of the people would have been, and, indeed, often was, furiously raised against the strangers who pretended to claim a portion of their subjects' allegiance, and preached the equal rights of all mankind incident to their universal brotherhood. But while these restless warriors were absent from their homes, or occupied in their debauches, the minister of Christ found his opportunity for insinuating his principles into the mind of the serf and the villain. The persecutions of which we read were mostly caused by the enmity of the great; the people, even in the regions of barbarism, were little disposed of themselves to quarrel with innocent men bent only upon doing them good.

The hostility, however, of the proud nobility was perhaps less injurious to the influence of the clergy than the favour into which they latterly received them; when these pious ministers had been led to follow their evil ways, to turn the gifts of the faithful into landed property, to found estates, and erect castles or palaces, as feudal lords themselves, the cause of the Church was then most truly in peril when the bishoprics became converted into benefices, to be conferred by the forms of feudal investiture at the hands of the temporal sovereign, and bound to furnish military service. The process by which this fatal endowment was effected was, no doubt, the natural, perhaps the inevitable, result of social and political circumstances. Year by year more and more of the land became absorbed by the clergy, into whose hands it was thrust, by grant or bequest, by their disciples and admirers. But the defence of the State depended upon the military service of all who held the land of the State in fee. The secular barons were required to do such service for the State; there seemed to be no excuse for the clergy in abandoning so plain a duty. Thus it was that the scandal arose, and became so widely extended, of the higher clergy assuming a secular, and even a military character; and losing in a great degree their proper position as ministers of peace and teachers of a religion which demands, above all things, the subordination of all worldly interest to the highest spiritual objects.

This peril had been foreseen, indeed, from an early period. The records of the Church show that measures were taken from the first to avert or modify it. Petitions were presented to Charlemagne to relieve the bishops from their military service, that they might devote themselves wholly to their spiritual charges, in which they could serve their feudal sovereign as well, or better, by prayers and works of charity. These views were upheld successively at councils, held, as we read, at Mayence, at Aix-la-Chapelle, at Augsburg; and they availed at last to control the despotic demands of the emperor, and the worldly pride of the bishops and clergy also.[1] The great episcopal principalities remained as a vestige of this early union of Church and State, and have hardly been entirely abolished even at the present day; but the character of the prelates was relieved from the scandal of bearing arms and of shedding blood.

It is probable that the severance of the clergy from the secular professions would hardly have been effected in the middle ages against the general temper of the times and the secular tendencies of human nature

[1] Ozanam, 'Études Germaniques,' ii. 293.

itself, had not the interests of the Papacy enforced upon them the law of celibacy. The head of the Church could not be blind to the imminent revolt of the clergy from his paramount authority, when they became admitted within the circle of the temporal nobility of the distant chiefs and sovereigns, who were already chafing under the spiritual restraints he claimed to lay upon them. It was of essential importance to the Pope that the clergy should be detached from the world by an insuperable barrier, and no means could be devised to effect this object so powerfully as the confirmation of the principle of clerical celibacy, which prevented the priest from marrying and generating children, from forming secular alliances and looking to the foundation of a family to succeed him. We shall hardly wrong the rulers of the Church in the middle ages, if we suppose that in the enforcement of celibacy, it was their main object to make use of the clergy as instruments of their own authority. It is true indeed that the principle was one of much earlier standing; it may date as a law and discipline from the fourth century; as an idea less distinctly viewed and more loosely held, it must be referred to a still earlier period. But in its original phase it bore no reference to political or ecclesiastical objects. The superstitious notion of the impurity of matter, the sinfulness of all fleshly indulgence, the necessity of ascetic mortification for the discipline of the body, the root of all evil, had made its way into the Church from the East, and from the contact of the Eastern Church with Manichean fanaticism. The multitude, at the same time, had been encouraged to view with a blind admiration the exercise of self-denial as in itself a virtue, without reference to the object to which it is directed; and thus the law of celibacy had been found to be an effective instrument in raising the estimation of the clergy in the eyes of their disciples. This potent instrument was still at hand, and the Church, under the fiat of the Roman pontiffs, now used it with increased severity to recover the clergy to the service of the Holy See, from which they seemed on the point of escaping. We may allow that the time was not yet ripe for a revolt from the Roman tyranny. The law of clerical celibacy had still an important part to play in confirming the influence of the missionary preachers over the simple people whom they were engaged in evangelizing. The Germans had been noted, even in the time of Tacitus, for their respect for purity in both sexes, and they would be more inclined to remark the restraints thus imposed upon themselves by the teachers of a Gospel of purity, which came to them from the South, the hotbed, as they might not unjustly imagine it, of a gross and voluptuous civilization. We must regard this law of celibacy, however licentiously it was evaded, and whatever were the evils inseparably connected with it, as an important agent in the great work of recommending Gospel truth to the wild and barbarous races of the North.

But this violation of natural instincts, whether for good or for evil, was confirmed by the practice of the monastic system, which received a fresh and powerful impulse at the beginning of the ninth century, a period, as we have seen, highly critical in the progress of the faith. The old rule of St.

Benedict of Nursia had reigned for three centuries, and diffused itself over the West of Christendom. It is to this rule that the German nation owe the early instructions of St. Gall, of Fulda, and of Corbey, which had inaugurated the missions into Suabia, Thuringia, and Saxony. " Three virtues constituted the sum of the Benedictine discipline : silence, with solitude and seclusion, humility and obedience, which, in the language of its laws, extended to impossibilities. All is thus concentrated on self. It was the man isolated from his kind who was to rise to a lonely perfection. All the social, all patriotic virtues were excluded: the mere mechanical observance of the rules of the brotherhood, or even the corporate spirit, is hardly worthy of notice, though they are the only substitutes for the rejected and proscribed pursuits of active life. The three occupations of life were the worship of God, reading, and manual labour. The adventitious advantages, and great they were, of these industrious agricultural settlements, were not contemplated by the founder. The object of the monks was not to make the wilderness blossom with fertility, to extend the arts and husbandry of civilized life into barbarous regions; it was solely to employ, in engrossing occupation, that portion of time which could not be devoted to worship and study."[1] Between these three duties, the hours of the day were strictly divided. Great abstinence was required, and the use of flesh and wine almost wholly forbidden. Not only was celibacy enforced, but the most innocent conver-

[1] Milman, ' Hist. of Latin Christianity,' book iii. ch. 6.

sation between the sexes forbidden. The full acceptance of these rules and restrictions exalted the reputation of the monks for sanctity. They were held to the public gaze as patterns of the highest Christian life, and continued for a long time to recommend the faith of Christ to the multitude by their example, whether they remained in their cells at home, or went forth to preach the Gospel among the people.

But this ancient rule, notwithstanding many bright exceptions, had degenerated from its original strictness, not without a corresponding loss of power and influence. It was restored and invigorated in the fashion of a later day, by another Benedict, born on the river Aniane, in the south of France, in the year 750. This well-meaning reformer added to the original rule of worship, study, and labour, a long code of minute injunctions on such lesser matters as clothes and diet. "Of eighty articles, twenty-one," says M. Guizot, "are foreign to any religious sentiment or moral intention. Nothing," he continues, "less resembles the gravity, the enthusiasm impressed upon the earlier rule of his predecessor; nothing attests more plainly the decay of the monastic spirit, and the rapid steps by which it was declining into a frivolous superstition. Like the monk of Nursia of old, the second Benedict desired the reform of monastic life; but the reform of the sixth century had been at the same time liberal and impassioned; it had addressed itself to all that was great and strong in human nature; that of the ninth century is puerile and subaltern in character, and addresses itself to all that is feeble and servile. And such was from this time forward the general character of the monastic institution; it lost its original grandeur and ardour, and remained laden with these frivolities, these absurd obligations which degrade men, even when they subject themselves to them in earnestness and devotion."[1]

It is necessary to insist upon this grave criticism from the pen of one so learned and so thoughtful, inasmuch as though the increased activity of the monastic system thus reformed imparted considerable strength to missionary efforts in the eyes of the heathens or weakly converts, on whom they were exerted, and constituted, in fact, an important factor in the propagation of the faith in its own day, the permanent effects it produced upon the character of Northern Christianity were in many respects disastrous. The labours of the Benedictines tended in the end to the corruption of the Church, which provoked and required the no less thorough reformation than that of Luther.

We will not bring this survey to a conclusion, however, without noticing one other influence which was brought to bear on the conversion of the Germans, and one to which, happily, no such stigma can attach. Even in the Middle Ages there were some intelligent learners among those who felt the power of Christian literature to open the sources of religious feeling. Indications are not wanting of the attention

[1] Guizot, 'Civilisation en France,' leçon xxvi. 159. Montalembert and others may be compared on the other side.

which was paid to education in the Church as early as the sixth century, implying some revival of the literary instinct after the first fury of the barbarian invasions. The decrees of councils may be cited from Gaul, Spain, Italy, and even Britain, of this and the age next following. The rule of the first Benedict expressly enjoins the practice of reading on Sundays and holydays as a part of the monastic discipline; and his monks seem to have taught as well as studied. It is recorded that they cherished a tradition that Varro, the most learned of the Romans, had resided at Monte Cassino, the cradle of their order, and held it as an incentive to that devotion to letters for which their name has been in later times so justly famous. From this time the list of schools established at the episcopal sees, particularly in France, becomes more and more numerous. The bishops themselves took in many instances an active part in literary instruction. Many monasteries became noted for the libraries they had collected, and for their devotion to the copying of ancient manuscripts. In the rude and bloody times of the Merovingian monarchy, we read with some surprise that a great school, both ecclesiastical and secular, was enshrined within the royal palace, under the direction of the sacred ministry. The young nobles of Frankish descent were collected together at these seminaries, as the youth of the German barbarians had been wont to congregate in the household of their chiefs; they there received an education in all the studies of the time, in company with the boys whom the king retained as hostages for the fidelity of the tribes he had subjugated. The study of Latin was one of the first elements in this education, and a considerable acquaintance was acquired with the literature of the ancient Romans, as well as with the writings of the doctors of the Church. The schools of Ireland, which flourished eminently through two centuries, taught not Latin only, but also Greek.

But the Carlovingian period, when the Church was making its most ardent efforts for the conversion of the Germans, was at the same time most distinguished for its application to education and literature. Charlemagne was himself eminent as a patron of letters. He brought back with him from Rome after his coronation, when he had arrived at the summit of his glory, the most able teachers of the age, and required them to revive or advance throughout his vast dominions all the sciences known to his age. The heathens of the North could not fail to be struck by the intellectual culture of the missionaries who came as preachers among them. The physical wonders which these strangers claimed as tokens of their divine mission, did less perhaps to exalt their reputation among their neophytes than the mental superiority which their early education had given them. We cannot doubt that the German intellect, with its natural taste, its curiosity, and its subtlety, was profoundly moved, even in these rude and simple ages, by the moral grandeur of a learned priesthood which condescended to enlighten its ignorance, to reason with its understanding, to refine its barbarism, and in short to open the mysteries of the heart and intellect to all who would receive with humility the spiritual doctrines revealed to them. We may be reminded of the readiness with which the Germans in the time of Augustus imbibed the lessons of their civilized adversaries at the court of the emperor or in the camp of his generals.

CHAPTER XI.

MORAL INFLUENCE OF GOSPEL TRUTH IN THE CONVERSION OF THE NORTHERN NATIONS.

We have reviewed the sequence of events which constitutes the history of the great Northern conversion, within the limits which are prescribed to one little volume. To this historical survey have been appended two more chapters, in which attention has been directed to the chief external causes which operated towards that conversion; namely, first, the influence of the secular authority which was enlisted in its favour; and secondly, that of the spiritual power assumed and exercised mainly by the Latin Church and the See of Rome. But another cause yet remains to be considered in the creeds and traditions, the moral habits and associations of the Northern nations themselves—in short, the internal influences which predisposed their minds to the reception of the Christian revelation. Upon this interesting subject a few remarks may be added in conclusion.

The general character of the religion common to the German nations in the time of Cæsar and Tacitus has been briefly noticed in our opening chapter. Those ancient authors saw but little of it below the surface, and that little they regarded, perhaps, through a distorted medium; nor do the subsequent writers of the period of the conversion add anything that is important for our fuller understanding of it. But we may reasonably conjecture that the religious ideas of the German races between the Danube and the Baltic coincided mainly with those which prevailed among the Scandinavians still further north, a people of kindred stock, language, and habits. The Teutonic inhabitants of Sweden and Denmark retained their ancient heathenism as late as the eleventh century. It so happens, by a curious coincidence, that at the last moment, when they were on the point of abandoning the faith of their ancestors, the patriotic spirit of one Sæmund, apparently himself a Christian priest, impelled him to collect the perishing traditions of his people, and preserve them for the respectful regard of posterity, even while it rejected their claim to belief and reverence. This collection is known by the name of the Edda, or " the Grandmother," reminding us of the *Veteres Aviæ*, the term applied to dear but exploded superstitions by the philosophic poet of ancient Rome. These traditions were surely still dear to the heart of the new converted Christian, who had openly renounced them ; perhaps he fondly recognised in them a germ of true belief, such as might form a basis for the intelligent acceptance of Church principles among his still hesitating countrymen.

It would seem from the Edda that the dominating idea of the Teutonic religion was the eternal conflict between good and evil. The good principle is constantly represented therein as all-powerful, as the Creator of the gods, as itself God. Nevertheless this principle is thwarted, and from time to time overcome by the opposite principle of Evil. Again and again the Evil One is baffled; but the final solution of the struggle is deferred from age to age, and remains to the present day, promised only, but never yet effected. To this Principle or Person no name is assigned; but He is adored under the mystery of a Divine Trinity, which is designated trebly as the "Lofty One," the "Equally Lofty," and the "Third One." Of this exalted Being, who governs all things, it is asserted that He will come hereafter to judge the world; to His power there will then be no effectual hindrance; His supreme judgment will be executed at a place removed from the great conflagration of the universe. Then, it is added, as if the picture were already present in the mind of one who knows no distinction of time, the spirits of the good inhabit an abode more brilliant than the sun; but the wicked shall be removed from the light of day, to the shores of the dead, to the sad dwelling where the wolf tears, and the worm devours them. From this transcendental theology we are soon drawn down to a lower level. We read of the generations of the gods impersonated in Odin and his twelve assistant divinities; of the giant Ymir, who sustains the part of the evil principle, and of the wicked race of giants proceeding from him; of the combat incessantly renewed between these spiritual potentates; of the nine worlds, in the centre of which is placed our earth; of Ygdravil, the sacred ash-tree, and the Nornas or Fates, who sit spinning men's destinies at its roots. To these may be added other marvellous legends, some of which may be traced in the Jewish Scriptures, while others reappear, again and again, in the worldwide superstitions of every continent. Among the most striking of these is the tradition of the holy city, Asgard, with its temple in the midst, and its thrones for Odin and his twelve Ases. To Odin, or the All-father, is attached a second Trinity of Thor the Thunderer, Tyr the Lightener, and Freyr the Spirit of Production, the giver of fruits and harvests, of peace and plenty. Again, another æon or generation of divinities supervene. Hertha, the Earth, and Freya, Love, are enumerated among the beneficent powers; while from the race of giants springs Loki, a second principle of Evil, with Hela, Death, the dog Fenris, and the Great Serpent. Then follows the long and complex legend of Balder, which represents, in an imaginative and pensive strain, the impersonation of the Best and Fairest of Beings, in whose presence nothing impure, nothing unjust, can exist; but who is doomed by a mysterious destiny to suffer dissolution. All the Universe, moved by the prayers of his divine mother, combines to effect the charm which shall redeem him; but one evil spirit alone withholds his sympathy, and the combined effort of all the rest is baffled. Thenceforth all Evil is unchained upon the world; an age of iron, of wars and murders follows. Loki and Fenris revel unrestrained. Then rises Odin, and buckles on his armour, surrounded by the Ases and Alfes, his shining warriors. In the contest that ensues, Odin himself is swallowed by the dog Fenris. Evil still triumphs, not to be put down by any impersonation of the Good. Nevertheless a brighter sun shall hereafter arise. Nature shall finally gain the ascendency;

a new race of human beings shall spring from the wreck of all mankind; all evil shall pass away,—Behold a new Heavens and a new Earth! and Balder himself, the best of beings, revived and triumphant over all.

The mythological traditions preserved in the Scandinavian Edda, and particularly the legend of Balder, present us with the most detailed, and accordingly with the most picturesque, records of the early religion of the North. It is on this account mainly that we may think it worth while to refer to them here; for we are not in a position to assert that they are generic in their character, or that they actually contain the most primitive elements of a creed at any time universally received among the various races of the great Teutonic stock. But, however this may be, they deserve our consideration on a further account, inasmuch as they seem to make some approach, at least, to the peculiar doctrines of Christianity, and have been supposed, accordingly, to favour the reception of the true faith among the races of the North. The reader will judge for himself from this review how far the Teutonic ideas of religion did actually approximate to those of our Revelation. That there is, indeed, a curious analogy between the two on certain points will be apparent to all candid inquirers; such as the pre-eminence of a sole invisible unknown author of the world, a spiritual being whom no image can personify, no temple can contain; the triple nature assumed in another phase by the same deity: the personal conflict between the good and the evil principle, with its alternate victories reflected in the fortunes of mankind; and the final triumph yet to be accomplished of the good over the evil. We may remark how the particular incidents in Scripture of Adam's sleep, the tree of life, the serpents, the flood, the ark, the rainbow of promise, have their corresponding legends in the Edda. But, most striking of all, the destiny of the world is attached in the rude imagination of the Northmen to the death of a God-man, who dies to revive triumphant, and the great drama of the world's history terminates in a final judgment of reward and punishment. We may fairly assume that such harmonies and analogies as these· did actually contribute to recommend the teaching of the Church to the Teutons, and smoothed away the inveterate prejudices against it. But we shall remark, no doubt, that most of these points of mutual resemblance are reproduced, more or less nearly, in other ancient mythologies, such as the Hindoo, the Persian, the Greek, and the Roman; and again, that none of them touch upon the really vital and distinctive characteristics of our faith,—the sinfulness of man's nature, and his redemption from sin by God's love and the self-sacrifice of the Divine Saviour.

But it is not perhaps to the actual religious traditions of the Northern nations that we should look for the basis of a reasonable reception of Christian truth among them, but rather in the moral and spiritual elements which pervade their whole religious system. We may fairly assert that to the conception of these races, the life of man appeared as a state of conflict divinely appointed. Odin, the greatest of the gods and universal fathers, maintains his eternal contest with the giants; but this contest assumed in their eyes a moral significance. Man was ordained, in their views, to take his own part in this combat. His destiny was not only to conquer the rude powers of nature surrounding and oppressing him, and to reduce them to vassalage, but he must overcome the gift of sensual passion, which he too has received from the evil spirit. His business through life was to fight and struggle. Hence, it may be said, the whole career of the German nations acquired its combative character. Odin, the spirit of moral, no less than of physical force, becomes pre-eminently the God of War. It is he that animates the courage of the warrior in the battle, he that restores him to life and receives him in the sacred halls of Valhalla. We may conceive how readily the Goths and Germans, with this conception of the Deity and of human life, might embrace the figurative representations in Scripture of God as the Lord of Hosts, and Christ as riding forth to battle. We may realize the struggle in the mind of the good Ulphila, when he debated whether to suppress the Book of Kings in his version of the Divine Word, as lending a too potent and direct encouragement to the carnal, rather than the spiritual conception of the Divine nature. But to the German as well as to the Jew, God was above all and before all, a Spirit. However the higher view of his nature might, from time to time, be clouded, his instinct ever led him back to the conviction that God was the Lord of spiritual combat, the Power which overthrew the evil tendencies of the human heart ; the conqueror, first of the evil within it, and then of the personal Evil One external to it. In this sense, the doctrines of the lowly and peaceable Jesus might be received unconsciously into the breasts of the Northern warriors, and the good seed once admitted might quickly grow and flourish.

The special honour in which the female sex was held, as it would seem, throughout the races of the North, has been often referred to as an important aid to the propagation of the Gospel among them. In nothing, certainly, were these people more plainly distinguished, by the testimony of Tacitus, our gravest authority, from the Greek and Roman pagans. This chivalrous feeling of respect towards women, which forms so marked an element of Teutonic life, attested not by the statement of an early writer only, but by the laws of later generations, has been traced by modern antiquarians to the notion familiar to the primitive Teutons, of a close intercourse between the flower of their heroes and the superior female existences, to which they gave the name of Walkyren, the heavenly virgins who received them, after a warrior's death, in the paradise of Valhalla. Under the influence, as we may suppose, of this soothing creed, the Germans held their women in esteem, and even in awe. The women, thus exalted in the eyes of their male associates, became their partners in the highest concerns of life, attending upon them in council, and admitted to a share in their hopes and fears and spiritual aspirations. The same process of true human culture was already in progress among all the disciples of the Christian faith in every clime. In the south of Europe, in Asia and in Africa, it was marking a distinction between the follower of Christ and the most earnest devotee of the most spiritual of pagan creeds. The German especially could not fail to recognise in the Gospel, which exalted woman in the person of the Virgin Mother of Christ, a principle congenial to his own natural feelings.

While paying this signal honour to women, the Germans exacted from them in return the strictest regard to their own purity. The duty which they thus demanded of their companions, they accepted with gallant devotion for themselves. Sins against chastity they did not hesitate to denounce, in the case of the one sex as well as of the other. Whatever laxity and corruption might still be justly imputed to them, they were, on the whole, famous in their generation for their superior moral virtues in this particular. Such was the testimony of Tacitus in the second century, and such was the testimony of writers even under the Later Empire, who declared, even of the Goths and Vandals, that they were not only distinguished for purity themselves, but exercised a marked effect in purifying the vitiated character of the Romans they conquered. It was from the fourth century that the grace of chastity acquired its highest exaltation in the teaching of the Church; and we may suppose that the honour thus done to the discipline they most esteemed, served in no mean degree to recommend its dogmatic teaching.

We may add further that the leading principle of the Christian faith, the personal relation and responsibility of every man to his Maker, was specially adapted to seize on the imagination of the Germans. The people of Southern Europe generally regarded their relation to the Deity as national rather than personal. Their God was the Author, or Judge, or Defender of the nation, not of the individual. Men fancied themselves living under a public and social compact with Him, and their private virtues or vices as of little concern to Him. The man was lost in the State. In the Christian scheme of individual salvation such ideas are entirely reversed, but among the Germans there was little or no sense of national unity. They possessed no cities to foster the idea of civil or municipal obligations to the Supreme Ruler. In their villages, as in their forests, every man lived for himself only, or combined occasionally with his fellow for the gratification of a sudden impulse to the war or to the chase. Here, then, was another point in which the Gospel might recommend itself to the Teutonic instinct, and secure acceptance with it. Nor can we doubt that to minds thus predisposed to set the individual before the nation, the cardinal Christian doctrine of immortality was supreme in interest; for it might seem to set the seal to the principle of each man's personal worth in the eyes of his Creator as something far transcending the importance of the city or the state, or the whole race and nation. The heathens of the North had already very generally embraced a vague idea of existence hereafter, and this idea the doctrine of the resurrection at once defined and deepened to them. The teaching of the mediæval Church, with its gross material pictures of its heaven and its hell, made a profound impression upon them, and in no portion of Christendom were these sensuous conceptions of spiritual things more promptly embraced, or more profusely illustrated, by ever-growing legends and traditions than in the Germany of the Middle Ages.

From these considerations we shall readily allow that there were some moral and religious notions current among the Teutonic races, such as might naturally predispose them to the reception of the faith in Christ. But we cannot doubt, or rather we would jealously maintain, that all mankind, of every race and origin, have a common interest in Gospel truth, and, in the most important respects, have some common elements akin to it. The love of God touches every heart of man; the power of God demands every man's attention; the justice of God warns and alarms all men. The spectacle of general corruption, such as had befallen the pagans of the South at the time of our Lord's appearance, could not fail to shock the natural sense of right with which men are everywhere born into the world. Doubtless the advent of Christ was providentially timed so as to seize the moment of the reaction against pagan abominations which had already set in under the purer, but still uncertain, teaching of the later philosophy. It was at this critical period, when the better spirit of human nature was thus darkly and almost hopelessly striving against the power of evil, that the hand of God was slowly fashioning the Teutonic nations to become His instruments for the propagation of His true faith in the remote corner of the world where they had been faintly descried' by the infidel and the pagan, by a Cæsar and a Tacitus.

Society for Promoting Christian Knowledge.

THE FATHERS FOR ENGLISH READERS.

Fcap. 8vo., Cloth boards, price 2s. each.

SAINT AUGUSTINE.
By the Rev. WILLIAM R. CLARK, M.A., Prebendary of Wells and Vicar of Taunton.

SAINT JEROME.
By the Rev. EDWARD L. CUTTS, B.A., author of "Turning Points of Church History," &c.

THE APOSTOLIC FATHERS.
By the Rev. H. S. HOLLAND, Student of Christ Church, Oxford.

THE DEFENDERS OF THE FAITH; or, The Christian Apologists of the Second and Third Centuries.
By the Rev. F. WATSON, M.A., Rector of Starston, Norfolk.

CONVERSION OF THE WEST.

Fcap. 8vo., Cloth boards, price 2s. each.

THE CONTINENTAL TEUTONS.

By the Very Rev. CHARLES MERIVALE, D.D., D.C.L., Dean of Ely. With Map.

THE CELTS.

By the Rev. G. F. MACLEAR, D.D., Headmaster of King's College Schools. With two Maps.

THE ENGLISH.

By the same Author. With two Maps.

THE NORTHMEN.

By the same Author. With Map.

THE HOME LIBRARY.

Crown 8vo., Cloth boards, price 3s. 6d. each.

THE HOUSE OF GOD THE HOME OF MAN.

By the Rev. G. E. JELF, M.A., Vicar of Saffron Walden.

THE INNER LIFE, as Revealed in the Correspondence of Celebrated Christians.

Edited by the late Rev. T. ERSKINE.

SAVONAROLA: his Life and Times.

By the Rev. WILLIAM R. CLARK, M.A., Prebendary of Wells and Vicar of Taunton, author of "The Comforter," &c.

NON-CHRISTIAN RELIGIOUS SYSTEMS.

Fcap. 8vo., Cloth boards, price 2s. 6d. each.

BUDDHISM.
By T. W. RHYS DAVIDS, of the Middle Temple. With Map.

HINDUISM.
By MONIER WILLIAMS, M.A., D.C.L., &c. With Map.

ISLAM AND ITS FOUNDER.
By J. W. H. STOBART, B.A., Principal, La Martinière College, Lucknow. With Map.

THE CORAN : its Composition and Teaching, and the Testimony it Bears to the Holy Scriptures.
By Sir WILLIAM MUIR, K.C.S.I., LL.D.

THE HEATHEN WORLD AND ST. PAUL.

Fcap. 8vo., Cloth boards, price 2s. each, with Map.

ST. PAUL IN DAMASCUS AND ARABIA.
By the Rev. GEORGE RAWLINSON, M.A., Canon of Canterbury, Camden Professor of Ancient History, Oxford.

ST. PAUL IN GREECE.
By the Rev. G. S. DAVIES, M.A., Charterhouse, Godalming.

ST. PAUL AT ROME.
By the Very Rev. CHARLES MERIVALE, D.D., D.C.L., Dean of Ely.

ST. PAUL IN ASIA MINOR, AND AT THE SYRIAN ANTIOCH.
By the Rev. E. H. PLUMPTRE, D.D., Prebendary of St. Paul's, Vicar of Bickley, Kent, and Professor of New Testament Exegesis in King's College, London.

DEPOSITORIES :
77, GREAT QUEEN STREET, LINCOLN'S-INN FIELDS, W.C. ;
4, ROYAL EXCHANGE, E.C. ; AND 48, PICCADILLY, W.

www.ingramcontent.com/pod-product-compliance
Lightning Source LLC
Chambersburg PA
CBHW030820190426
43197CB00036B/636